JIM TOTTEN'S PRBLEMS OF THE WEEK

JIM TOTTEN'S PROBLEMS OF THE WEEK

Editors

John Grant McLoughlin
University of New Brunswick, Canada

Joseph Khoury
University of Ottawa, Canada

Bruce Shawyer
Memorial University of Newfoundland, Canada

World Scientific

NEW JERSEY · LONDON · SINGAPORE · BEIJING · SHANGHAI · HONG KONG · TAIPEI · CHENNAI

Published by

World Scientific Publishing Co. Pte. Ltd.

5 Toh Tuck Link, Singapore 596224

USA office: 27 Warren Street, Suite 401-402, Hackensack, NJ 07601

UK office: 57 Shelton Street, Covent Garden, London WC2H 9HE

British Library Cataloguing-in-Publication Data

A catalogue record for this book is available from the British Library.

JIM TOTTEN'S PROBLEMS OF THE WEEK

ISBN 978-981-4513-30-2

Printed in Singapore

Dedication

James Edward Totten, 1947-2008

With the sudden death of Jim (James Edward Totten) on March 9, 2008 the Mathematics Community lost someone who was dedicated to mathematics education and to mathematical outreach. Jim was born August 9, 1947 in Saskatoon and raised in Regina. After obtaining his Bachelor's degree at the University of Saskatchewan, Jim then earned a Master's and PhD degree in Mathematics from the University of Waterloo, after which he joined the faculty at Saint Mary's University in Halifax. Robert Woodrow first got to know Jim when he and Jim shared an office at the University of Saskatchewan while Jim visited there during 1978-1979. That was a long cold winter, but Jim's active interest and enthusiasm for mathematics and the teaching of mathematics made the year a memorable one. The next year Jim took up a position at Cariboo College, where he remained as it evolved into the University College of the Cariboo and then eventually into Thompson Rivers University, retiring as Professor Emeritus in 2007.

During his years in Kamloops, Jim was a mainstay of the Cariboo Contest, an annual event which brought students in to the college and which featured a keynote speaker, often drawn from Jim's list of mathematical friends. This once included an invitation to Robert, which featured a talk on public key encryption mostly memorable for the failure of technology at a key moment, much to Jim's amusement.

Jim became a member of the CMS in 1981, and joined the editorial board of Crux in 1994. When Bruce was looking for someone to succeed him as Editor-in-Chief, there was no doubt in his mind whom to approach. Bruce Shawyer spent a week in Kamloops staying at Jim's home and working with Jim and Bruce Crofoot to smooth the transition. Jim's attention to detail and care was appreciated by all, particularly those contributing copy that was carefully checked, as Robert Woodrow gladly confirms from his continued association with Jim through the Olympiad Corner, an association which continued to the end.

Jim loved his Oldtimers' hockey and was an avid golfer. When not playing hockey or golf, he was active with the Kamloops Outdoors Club. Jim was never just a participant, always an active volunteer.

Jim is survived by his loving wife of 40 years, Lynne, his son Dean, daughter-in-law Christie and granddaughter Mikayla of Sechelt, his father Wilf Totten of Edmonton, sister Judy Totten of Regina, sister Josie Laing (Neil) of Onoway, sister-in-law Marilyn Totten of Regina, sister-in-law Constance Ladell (David Dahl) of Kamloops, many cousins, family friends and mathematical colleagues. All miss his warmth and love.

James Edward Totten, 1947-2008
Photo by Don DesBrisay

Preface

James (Jim) Totten began his university teaching career at Saint Mary's University in Halifax, Nova Scotia in 1976. There he started his Problem of the Week initiative. Typically a problem would be posted for undergraduate students to solve. The success and positive feedback led Jim to continue with this for over 30 years. Jim Totten taught in Kamloops, British Columbia from 1979 until his retirement in 2007. The institution evolved from Cariboo College to University College of the Cariboo and subsequently, Thompson Rivers University (TRU). Each week of the fall and winter semesters featured a problem different from all others preceding it until the final year or two when Jim took the liberty to reuse some personal favourites. Jim compiled a collection of eighty problems used up to 1986 in *Problems of the Week*, Volume 7 of the ATOM (A Taste of Mathematics) Series published by the Canadian Mathematical Society. None of these problems are included in this book.

Jim Totten passed away suddenly in March 2008. Two of Jim's colleagues, namely, Fae deBeck and Rick Brewster, along with myself (John Grant McLoughlin) coordinated organization of a conference aptly named *Sharing Mathematics: A Tribute to Jim Totten*. A gift of two large red binders was bestowed upon me during this process. Alongside came encouragement to bring this collection to others so that it would be enjoyed widely.

Jim never pretended that the problems were original. The problems come from many sources, including several brought to his attention during his graduate studies at University of Waterloo from 1968 to 1974. It was there that Jim became acquainted with Ross Honsberger. Jim described himself as a *willing listener* when Ross wanted to share interesting problems or solutions with someone. This excitement for gems was contagious to Jim, and he proceeded to carry forth his own love of problems with a commitment to sharing that spirit of his own.

It is in Jim's spirit that this collection is being shared with readers and problem solvers of all sorts. The problems in this book are generally accessible to strong high school students though specifically they were intended for undergraduate students. Many of the problems have a recreational or puzzling flavour, in that Jim was one who dearly enjoyed such challenges. Such problems tend to attract broader levels of interest than specific topics like trigonometry or calculus may do. The collection here is intended to attract many, thus, providing entry points for anyone interested in problem solving. Enjoy the challenges.

John Grant McLoughlin
University of New Brunswick

Jim Totten served as a Problems Editor on the board of Crux Mathematicorum for many years, before becoming Editor-in-Chief.

We are very grateful to have been given access to this collection and have great pleasure in presenting it to the wider mathematical community.

Joseph Khoury
University of Ottawa

Bruce Shawyer
Memorial University of Newfoundland

Contents

Chapter 1

Combinatorial Geometry

(1) *Problem.* Find the dimensions of all rectangles of size $m \times n$ which contain exactly 100 squares of all sizes, where each square has sides parallel to the edges of the given rectangle and has its corners at the grid points of the interior or boundary of the rectangle.

For example, a 2×3 rectangle would have 2 squares of size 2 and 6 squares of size 1, for a total of 8 squares.

Solution. The number of squares of a given size can be counted by considering the location of the upper right corner in the underlying rectangular grid. Let us assume that the smaller dimension of the underlying grid is n and that the other dimension is $n + k$.

The number of squares of size n (the largest possible) is $k + 1$; the number of squares of size $n - 1$ is $2(k + 2)$; in general, the number of squares of size $n - i$ is $(i + 1)(k + i + 1)$. Thus, the total number of squares is given by

$$\begin{aligned}
S &= 1(k + 1) + 2(k + 2) + 3(k + 3) + \cdots + n(k + n) \\
&= k(1 + 2 + 3 + \cdots + n) + (1 + 4 + 9 + \cdots + n^2) \\
&= \frac{n(n + 1)k}{2} + \frac{n(n + 1)(2n + 1)}{6} \\
&= \frac{n(n + 1)(3k + 2n + 1)}{6}.
\end{aligned}$$

We wish to solve for $S = 100$, which is the same as solving for

$$n(n + 1)(3k + 2n + 1) = 600.$$

Clearly n must be less than 8, and the values $n = 6$ and $n = 7$ both require that 7 divides evenly into 600.

Thus, we must only try the values of n from 1 to 5. Simple calculation shows that the only values of n which allow a solution are

$n = 1$, $n = 4$ and $n = 5$, for which the corresponding values of k are 99, 7, and 3 respectively.

Therefore, the dimensions of the only three rectangles meeting the requirements are:

$$1 \times 99, \qquad 4 \times 11, \qquad \text{and} \qquad 5 \times 8.$$

(2) *Problem.*   Find all different right-angled triangles with all sides of integral length whose areas equal their perimeters.

Solution.   Let the two legs have lengths x and y and let the hypotenuse be z. Then

$$x^2 + y^2 = z^2 \qquad \text{and} \qquad \frac{xy}{2} = x + y + z.$$

This gives

$$x^2 + y^2 = \left(\frac{xy}{2} - x - y\right)^2 = \frac{x^2 y^2}{4} + x^2 + y^2 - x^2 y - xy^2 + 2xy.$$

Simplifying, we obtain $xy(xy - 4x - 4y + 8) = 0$.

Since we rule out degenerate triangles, we get $xy - 4x - 4y + 8 = 0$ and thus, $(x - 4)(y - 4) = 8$.

Let us now suppose that x is the larger of the two values. Since x and y are integer values only, either we have $x - 4 = 8$, $y - 4 = 1$, which yields $x = 12$, $y = 5$ and $z = 13$, or we have $x - 4 = 4$, $y - 4 = 2$, which yields $x = 8$, $y = 6$ and $z = 10$. Therefore, the only triangles with the given properties are the 5, 12, 13 and 6, 8, 10 right-angled triangles.

(3) *Problem.*   Some of the squares on an infinite piece of (Cartesian) graph paper are coloured red, and the remainder are coloured blue. This is done in such a way that any 2×3 rectangle made up of six squares contains exactly two red squares. How many red squares may be included in a 9×11 rectangle made up of 99 squares?

Solution.   Let us coordinatize the graph paper with the centre of each square receiving integer-valued x- and y-coordinates. Suppose first that two squares with a common boundary are both coloured red, say $(0, 0)$ and $(1, 0)$. By considering the 2×3 rectangle bounded by $(0, 0)$, $(1, 0)$, $(1, 2)$, and $(0, 2)$, we see that the squares $(0, 1)$, $(1, 1)$, $(0, 2)$, and $(1, 2)$ are all coloured blue.

Now by considering the 2×3 rectangle bounded by $(0, 1)$, $(2, 1)$, $(2, 2)$, and $(0, 2)$, we observe that both $(2, 1)$ and $(2, 2)$ must be coloured red. But this implies that the 2×3 rectangle bounded by $(1, 0)$, $(2, 0)$, $(2, 2)$, and $(1,2)$ has at least three squares coloured

red, which is a contradiction. Therefore, no two red squares can share a boundary.

Let us impose our coordinatization of the graph paper such that $(0,0)$ is coloured red. Then $(0,1)$, $(1,0)$, $(-1,0)$, and $(0,-1)$ are all coloured blue. By considering the 2×3 rectangle bounded by $(-1,0)$, $(1,0)$, $(1,1)$, and $(-1,1)$, we see that exactly one of $(-1,1)$ and $(1,1)$ are coloured red, say $(1,1)$ is coloured red and $(-1,1)$ is coloured blue. Then, by considering the 2×3 rectangle bounded by $(0,0)$, $(1,0)$, $(1,2)$, and $(0,2)$, we see that both $(0,2)$ and $(1,2)$ are coloured blue, and by considering the 2×3 rectangle bounded by $(0,1)$, $(2,1)$, $(2,2)$, and $(0,2)$, we see that $(2,2)$ is coloured red and $(0,2)$ is coloured blue.

Continuing in this manner we can see that the squares (n,n) are coloured red for any integer n. It can also be shown that all squares $(n, n+3k)$ are coloured red for any integers n and k, and that all other squares are coloured blue.

This essentially says that every third diagonal strip is coloured red with the remainder coloured blue.

Now in a 9×11 rectangle positioned on the graph paper and containing 99 squares, it is readily seen that exactly 33 of them are coloured red, since when we simply up the number of squares in every third diagonal strip, we get the same value of 33 no matter which of the first three diagonals we begin with.

(4) *Problem.* Consider arrangements of pennies in rows in which the pennies in any row are contiguous, and each penny not in the bottom row touches two pennies in the row below.

For example, is allowed, but is not.

How many arrangements are there with n pennies in the bottom row? To illustrate, there are 5 arrangements with $n = 3$, namely

Solution. The desired number of arrangements takes on alternate values from the Fibonacci sequence: this sequence starts with two 1s, and each succeeding value is obtained by adding the previous

two values – thus,

$$1, 1, 2, 3, 5, 8, 13, \ldots.$$

The values we seek then are 1, 2, 5, 13, 34, 89, ...
To prove this we will work with two sequences. Let

a_n = the number of legal arrangements with n pennies in the
bottom row (this is the desired function), and

b_n = the number of legal arrangements with **at most** n pennies
in the bottom row.

Clearly

$$b_n = a_n + b_{n-1},$$

since an arrangement with at most n pennies on the bottom row
either has n pennies there or fewer.

Now consider the second row from the bottom in each of the patterns counted by a_{n+1}. If there is a penny in the left-most position
here, there are b_n arrangements for the second row up, by placing
each of the b_n patterns on top, left-justified. If the left-most position of the second row is vacant, then there are a_n arrangements:
for each pattern in a_n, add a penny to the left of the bottom row.
Thus we see that

$$a_{n+1} = b_n + a_n.$$

From these equations and the fact that $a_1 = b_1 = 1$, we conclude
that the sequence $a_1, b_1, a_2, b_2, \ldots$ is the Fibonacci sequence, and
the initial statement in our proof holds.

(5) *Problem.* Given eight small cubes—two each in red, blue, green,
and yellow—in how many different ways can they be assembled to
form a large cube so that each face of the large cube shows all four
colours? (Note: two large cubes are considered identical if one of
them can be turned so that the position of the colours on it matches
the position of those on the other.)

Solution. Let us first consider any single face. Since two colour
schemes are identical if one can be rotated so that it matches the
other, there are precisely 6 ways that the four colours can be rearranged on the face: we can (without loss of generality assume that
the upper left corner of the face is red; then there are 3 choices of
colour for the lower right corner; and for each such choice there are

2 ways to place the remaining 2 colours. Next observe that once the bottom four small cubes are in place, they determine completely the top four small cubes, since each remaining small cube must be placed in the opposite corner to the placement of the first small cube of that colour in order to appear on all six faces of the larger cube. Now if we start by placing the lower layer of small cubes in one of the six possible schemes, one can check that when the large cube is completed its six faces display all six of the allowable schemes. Thus, the large could be rotated to correspond to any starting position of the bottom four small cubes. Therefore, it is the only large cube that can be built, and there is exactly one such cube.

(6) *Problem.* Ewelme is a rather quiet village – except on the day of the annual sheep market. On that day, lines of hurdles are erected end-to-end from the flagpole to each of the corners of the marketplace (the marketplace is a quadrilateral, not necessarily convex) and along three of its four sides, making three triangular pens of different areas. Each hurdle is one metre long, there are no overlaps or gaps, and no hurdle is bent or broken. Also, the flagpole is erected within the perimeter of the marketplace.

Each pen is filled with sheep – one sheep to each square metre. The number of sheep in each pen is equal to the number of hurdles surrounding that pen.

What is the area of the Ewelme marketplace?

Solution. Let a, b, c be the sides of a triangle whose area is numerically equal to its perimeter. Let s denote its semi-perimeter; that is, $s = \frac{1}{2}(a + b + c)$. The perimeter of the triangle is $2s$, while, from Heron's Formula, its area is given by

$$\text{area} = \sqrt{s(s - a)(s - b)(s - c)}.$$

Setting these equal, squaring, and simplifying results in:

$$(s - a)(s - b)(s - c) = 4s. \tag{1}$$

In our problem the sides all have integer lengths. The above formula then implies that s must be an integer (at worst it would have been half of an odd integer, but this would leave the right side an integer and the left side not an integer). Let us set $x = s - a$, $y = s - b$, and $z = s - c$, all of which must be positive integers. Then,

$$x + y + z = 3s - (a + b + c) = 3s - 2s = s.$$

Equation (1) can then be rewritten as

$$xyz = 4(x + y + z). \tag{2}$$

Without any loss of generality we may assume that $x \leq y \leq z$, which is equivalent to $a \geq b \geq c$. Then

$$x^2 z \leq xyz = 4(x + y + z) \leq 4(3z) = 12z.$$

Therefore, $x \leq 3$.

Suppose that $x = 3$. Then equation (2) becomes $3yz = 4(3+y+z)$, or $3yz - 4y - 4z = 12$, which (when multiplied by 3) can be further rearranged to yield

$$(3y - 4)(3z - 4) = 52.$$

Since y must be an integer, we cannot have $3y - 4 = 1$, $3y - 4 = 4$, $3y - 4 = 13$, nor $3y - 4 = 52$. If $3y - 4 = 2$, then $y = 2$, but this contradicts $x \leq y$.

The only possibility left is $3y - 4 = 26$; but this means that $3z - 4 = 2$, and then $z > x$, another contradiction. Therefore, we are forced to conclude that there are no solutions with $x = 3$.

Suppose that $x = 2$. Then equation (2) becomes $2yz = 4(2+y+z)$, or $(y - 2)(z - 2) = 8$. In this case we must have $(y, z) = (3, 10)$ or $(y, z) = (4, 6)$.

The final possibility is $x = 1$. Then equation (2) becomes $yx = 4(1 + y + z)$, or $(y - 4)(z - 4) = 20$. In this case (y, z) must be one of $(5, 24)$, $(6, 14)$, or $(8, 9)$.

Since $s = x + y + z$, by putting the above together, we conclude that the only (integer) values for (a, b, c) are $(13, 12, 5)$, $(10, 8, 6)$, $(29, 25, 6)$, $(20, 15, 7)$, and $(17, 10, 9)$.

We must use exactly 3 of these 5 triangles to form the Ewelme marketplace. Let $ABCD$ be the marketplace and let F be the flagpole. Since adjacent triangles must have sides of the same length, the only triangles we need to consider are $(10, 8, 6)$, $(29, 25, 6)$, and $(17, 10, 9)$ with the common sides having lengths 6 and 10. There are only four possible configurations as shown on the next page:

The fourth possibility above may be ruled out since the flagpole F lies outside the quadrilateral. The quadrilateral's area can be computed by summing the areas of the 4 triangles ABF, BCF,

CDF, and DAF. The first 3 of these triangles have their areas equal in value to their perimeters. That is, the combined areas of ABF, BCF, and CDF is

$$(17+10+9)+(10+8+6)+(29+25+6) = 36+24+60 = 120 \text{ m}^2.$$

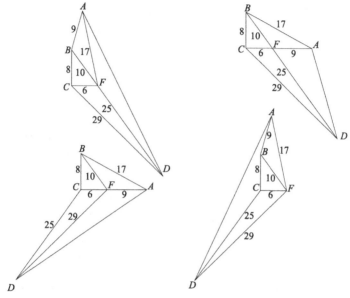

It remains to find the area of the fourth triangle DAF. In the second and third configurations above, we see that $\triangle BCF$ and $\triangle BCA$ are both right-angled with right angle at C since their sides satisfy the Pythagorean relation. Hence, C, F, and A are collinear. This means that $\triangle DAF$ and $\triangle CDF$ have the same altitude from the base line CA, and thus their areas are in the ratio as their respective base lengths. Therefore, $\triangle DAF$ has area $\frac{9}{6}$ times the area of $\triangle CDF$. That is, $\triangle DAF$ has area $\frac{3}{2} \cdot 60 = 90$ square metres. This yields a total area for the marketplace of $120+90 = 210$ square metres.

We will now show that also in the first configuration we have the area of $\triangle DAF$ equal to 90 square metres. To this end we must show that B, F and D are collinear. Then the area of $\triangle DAF$ will be $\frac{25}{10}$ times the area of $\triangle ABF$; that is, $\frac{5}{2} \cdot 36 = 90$ square metres, as claimed.

Extend the line BC beyond C to meet a line through D parallel to CF at a point E. By construction, angle CED is a right angle.

Drop a line parallel to CE from F meeting the line DE at G. Angle FGD is also a right angle. Let h be the length of the common altitude of $\triangle DCE$ and $\triangle DFG$. Let x be the length of the base of $\triangle DFG$. Then the length of the base of $\triangle DCE$ is $x + 6$. Using the Theorem of Pythagoras on the two newly-created triangles, we have

$$h^2 + x^2 = 25^2 = 625 \qquad (3)$$
$$h^2 + (x+6)^2 = 29^2$$
$$\text{or} \qquad h^2 + x^2 + 12x + 36 = 841. \qquad (4)$$

Subtracting (3) from (4) we get $12x + 36 = 216$, which yields $x = 15$. From (3) we compute $h = 20$. Now consider the right-angled triangle DEB. The length of its hypotenuse is

$$\sqrt{(8+20)^2 + (x+6)^2} = \sqrt{28^2 + 21^2} = 7\sqrt{4^2 + 3^2} = 7 \cdot 5 = 35.$$

Since the lengths of BF and FD sum to 35, we see that B, F, and D must be collinear, and we are finished.

(7) *Problem.*   Can 250 copies of a $1 \times 1 \times 4$ block be packed into a $10 \times 10 \times 10$ box?

Solution.   The $10 \times 10 \times 10$ cube contains 125 blocks of size $2 \times 2 \times 2$. Let us colour these blocks alternately black and white. Clearly, if we have a packing of the $10 \times 10 \times 10$ cube with $1 \times 1 \times 4$ blocks, any $1 \times 1 \times 4$ block will contain exactly two unit cubes of each colour. Thus, the packing is possible only if the box contains the same number of black and white unit cubes. However, there were 125 blocks which we coloured, and 125 is an odd number. Therefore, there cannot be the same number of black and white unit cubes, and any attempt to do the packing is doomed to failure.

Alternatively.   Divide the $10 \times 10 \times 10$ box into 125 $2 \times 2 \times 2$ portions; colour each portion black or white so that the whole box resembles a three-dimensional $5 \times 5 \times 5$ chessboard. Now 63 of the 125 portions are of one colour and 62 portions are of the other colour. Since each $1 \times 1 \times 4$ block must contain two unit cubes of each colour, the most blocks we can place in the box is $4 \times 62 = 248$. Thus, the task cannot be done.

(8) *Problem.*   The rectangle $ACEF$ is partitioned into an octagon $ABGHJDEF$ and a hexagon $BCDJHG$, as shown on the next page. The line segments DJ and GH are parallel to the side AC, while the line segments BG and HJ are parallel to the side CE.

The diagram (below) is <u>not</u> drawn to scale. The lengths of the sides of the octagon are 1, 2, 3, 4, 5, 6, 7, and 8 in some order. Assign the side lengths to the sides of the octagon in order to maximize the area of the hexagon. What is this maximum area?

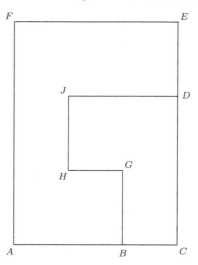

Solution. We first observe that $AF = BG + HJ + DE \geq 1 + 2 + 3 = 6$. Also, since we want to maximize the area of the hexagon and the side lengths assigned to BG, HJ, and DE can always be permuted without destroying the integrity of the diagram, we want DE to be the smallest of the 3 lengths.

Furthermore, we will always have a larger area for the hexagon if we ensure that $HJ > BG$ (to see this extend the line segment HG to meet CE and compute the areas of the two resulting rectangles which partition the hexagon).

Therefore, we need $DE < BG < HJ$, and $AF = DE + BG + HJ$. This leaves us with only four possibilities for the 4-tuple (DE, BG, HJ, AF):

$$(1, 2, 3, 6), \quad (1, 2, 4, 7), \quad (1, 2, 5, 8), \quad (1, 3, 4, 8).$$

Before considering these cases separately, we observe that

$$GH < DJ < EF \quad \text{and} \quad GH < AB < EF.$$

Again, to maximize the area of the hexagon, we want $DJ > AB$. Thus, $GH < AB < DJ < EF$.

Case (i): $(DE, BG, HJ, AF) = (1, 2, 3, 6)$.

Then $(GH, AB, DJ, EF) = (4, 5, 7, 8)$. Thus,

$$BC = DJ - GH = EF - AB = 3.$$

The hexagon then looks like:

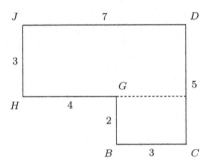

The area is clearly $3 \times 7 + 2 \times 3 = 27$.

Case (ii): $(DE, BG, HJ, AF) = (1, 2, 4, 7)$.
Then $(GH, AB, DJ, EF) = (3, 5, 6, 8)$. Thus,

$$BC = DJ - GH = EF - AB = 3.$$

The hexagon then looks like:

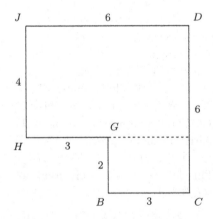

The area is clearly $4 \times 6 + 2 \times 3 = 30$.

Case (iii): $(DE, BG, HJ, AF) = (1, 2, 5, 8)$.
Then $(GH, AB, DJ, EF) = (3, 4, 6, 7)$. Thus,

$$BC = DJ - GH = EF - AB = 3.$$

The hexagon then looks like:

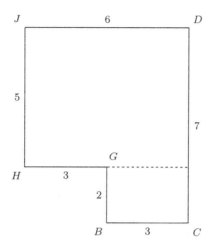

The area is clearly $5 \times 6 + 2 \times 3 = 36$.

Case (iv): $(DE, BG, HJ, AF) = (1, 3, 4, 8)$.

Then $(GH, AB, DJ, EF) = (2, 3, 6, 7)$. Thus,

$$BC = DJ - GH = EF - AB = 4.$$

The hexagon then looks like:

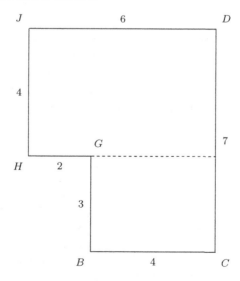

The area is clearly $4 \times 6 + 2 \times 3 = 30$.

The maximum occurs in Case (iii), and has the value 36. The assignment of lengths is as follows:

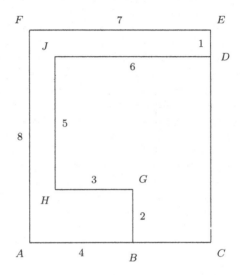

(9) *Problem.*   For what values of n can an $n \times n$ square be tiled using 2×2 squares and 3×3 squares? (By tiling the $n \times n$ square, we mean that its entire area is covered and no tile extends beyond the boundary of the $n \times n$ square.)

Solution. If n is evenly divisible by either 2 or 3, the tiling is clearly possible. In fact, we need only of the two possible tile types. We will now show that in all other cases the tiling is impossible.

Colour each square of an odd-numbered row of the $n \times n$ square black, and colour each square in an even-numbered row white. Let d be the difference between the number of small black and the number of small white squares that are visible. (As we continue to place tiles on the $n \times n$ square, the value of d will change.) Initially, the value of d is n, since we have assumed above that n is odd.

We now start placing the 2×2 and 3×3 tiles to cover up the small squares. A 2×2 tile covers up exactly 2 black and 2 white squares, and has no effect on d. Placing a 3×3 tile either lowers or raises the value of d by 3. Since d is NOT evenly divisible by 3, we can never reduce d to 0. This means that there are always some small squares visible, and therefore, a tiling is impossible.

(10) *Problem.*   On a $2n \times 2n$ board we place several non-overlapping $n \times 1$ polyominoes (each covering exactly n unit squares of the board) until no more $n \times 1$ polyominoes can be accommodated. What is the maximum number of unit squares on the board that can remain uncovered?

Solution. The following figure, which illustrates the case $n = 6$ and can be generalized, shows that the $2n \times 2n$ board can be *blocked* by $2n + 1$ of the $n \times 1$ polyominoes; that is, so that no further $n \times 1$ polyominoes can be accommodated.

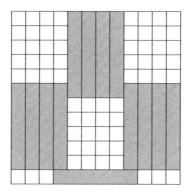

To show that $2n + 1$ is minimal, assume that the board is blocked by $2n$ polyominoes instead. We will derive a contradiction. Clearly, not all of them can have the same orientation (either horizontal or vertical).

It can also be seen that if, for any k with $k \leq n$, the kth row from the edge of the board contains a horizontally oriented $n \times 1$ polynomino, then so must the ith row from the same edge for every $i < k$.

Thus, any parallel rows containing a horizontally oriented polynomino must be consecutive from the nearest edge of the board. A similar observation can be made for vertically oriented $n \times 1$ polyominoes.

Suppose now that there are p rows which do not contain a horizontally oriented $n \times 1$ polynomino and q columns which do not contain a vertically oriented polynomino. Note that $p \geq 1$ and $q \geq 1$ by our initial observation above.

Thus, we must have a $p \times q$ rectangle which is totally uncovered. Since there are $2n$ polyominoes on the board and there a total of $4n$ rows and columns altogether, we see that $p + q \geq 2n$, which implies that either $p \geq n$ or $q \geq n$. In either case, the $p \times q$ rectangle can accomodate a further $n \times 1$ polyomino, which is a contradiction.

Thus, we may block further placement of $n \times 1$ polyominoes by using $2n + 1$ properly placed $n \times 1$ polyominoes.

In total, these $2n + 1$ polyominoes cover $(2n + 1)n = 2n^2 + n$ unit squares, leaving $(2n)^2 - (2n^2 + n) = 2n^2 - n$ uncovered, and this is the maximum number that can remain uncovered.

Chapter 2

Functions

(11) *Problem.* Consider the graphs of the functions $y = x^2$ and $y = a^x$ for $a > 1$.

Note that if $a = 2$, the graphs intersect at two distinct positive values of x, namely $x = 2$ and $x = 4$. On the other hand, if $a = 10$, the two graphs do not intersect for positive x. Clearly, there must be some value of a such that the graphs intersect for exactly one positive value (at this point the graphs will be tangent to each other).

What is this value of a? What is the value of x for which the graphs will be tangent to each other? That is, find the unique $a > 1$ such that $a^x = x^2$ has exactly one positive solution x, and find that solution x.

Can this be generalized to $a^x = x^n$?

Solution. We will consider the general case. That is, $y = a^x$ and $y = x^n$ are tangent for some value of $n \neq 0$. We have

$$a^x = x^n \tag{1}$$
$$a^x \ln a = n x^{n-1}, \tag{2}$$

where equation (2) is obtained by setting the derivatives of the two functions equal to each other. Note that x cannot equal 0. Substituting (1) into (2), we get

$$x^n \ln a = n x^{n-1}$$
$$\text{or} \quad x \ln a = n. \tag{3}$$

Taking (natural) logarithms on both sides of (1), we obtain

$$x \ln a = n \ln x. \tag{4}$$

15

We now combine (3) and (4) to obtain

$$n \ln x = n$$

$$\text{or} \qquad \ln x = 1$$

$$\therefore \qquad x = e.$$

But $a^x = x^n$ for this value of x. Thus, $a^e = e^n$. Consequently, $a = e^{n/e}$.

In the original problem, where $n = 2$, we have $a = e^{2/e}$.

(12) *Problem.*   How good is your understanding of functions and functional notation? Determine all strictly increasing functions f from the real numbers into the real numbers which satisfy

$$f(f(x) + y) = f(x + y) + f(0).$$

Solution.   First let us set $f(0) = C$. Then, the equation which must be satisfied for all x and y is

$$f(f(x) + y) = f(x + y) + C.$$

Now, if we let $y = 0$, we observe that

$$f(f(x)) = f(x) + C.$$

On the other hand, if we replace x by 0 and replace y by x, we get

$$f(C + x) = f(f(0) + x) = f(x) + C.$$

Since f is strictly increasing, for any given real value, say k, there can only be one real value s such that $f(s) = k$.

Therefore, since $f(f(x)) = f(C+x)$, we must conclude that $f(x) = x+C$. Thus, the only functions which satisfy the above "functional equation" are the functions $f(x) = x + C$ for some real constant C.

(13) *Problem.*   For each positive integer n, let

$$f(n) = \frac{1}{\sqrt[3]{n^2 + 2n + 1} + \sqrt[3]{n^2 - 1} + \sqrt[3]{n^2 - 2n + 1}}.$$

Determine the value of $f(1) + f(3) + f(5) + \cdots + f(999997) + f(999999)$.

Solution.   Let $a = \sqrt[3]{n+1}$ and $b = \sqrt[3]{n-1}$. Then

$$f(n) = \frac{1}{a^2 + ab + b^2} = \frac{a - b}{a^3 - b^3} = \frac{a - b}{2} = \frac{1}{2}(\sqrt[3]{n+1} - \sqrt[3]{n-1}).$$

Telescoping we get for all positive integers k:

$$\sum_{n=1}^{k} f(2n-1) = \frac{1}{2} \sum_{n=1}^{k} (\sqrt[3]{2n} - \sqrt[3]{2n-2}) = \frac{1}{2} \sqrt[3]{2k}.$$

In particular with $k = 500000$ we get

$$f(1) + f(3) + f(5) + \cdots + f(999999) = \frac{1}{2} \sqrt[3]{10^6} = 50.$$

(14) *Problem.* Determine all functions f from the non-zero integers into the rationals satisfying the following functional equation:

$$f\left(\frac{x+y}{3}\right) = \frac{f(x) + f(y)}{2}.$$

Solution. We will show that f is a constant (rational valued) function. First note that any constant (rational valued) function satisfies the functional equation. Next we note that:

$$f(1) = f\left(\frac{1+2}{3}\right)$$

$$= \frac{f(1) + f(2)}{2} \Rightarrow f(1) = f(2),$$

$$f(2) = f\left(\frac{3+3}{3}\right) = \frac{f(3) + f(3)}{2} = f(3).$$

Now suppose that for $k \geq 1$ we have $f(1) = f(2) = \cdots = f(3k)$. Then since $3k \geq k+2$ whenever $k \geq 1$, we have

$$f(2) = f(k+1) = f\left(\frac{3k+1+2}{3}\right)$$

$$= \frac{f(3k+1) + f(2)}{2} \Rightarrow f(2) = f(3k+1),$$

$$f(1) = f(k+1) = f\left(\frac{3k+2+1}{3}\right)$$

$$= \frac{f(3k+2) + f(1)}{2} \Rightarrow f(1) = f(3k+2),$$

$$f(3) = f(k+2) = f\left(\frac{3k+3+3}{3}\right)$$

$$= \frac{f(3k+3) + f(3)}{2} \Rightarrow f(3) = f(3k+3).$$

Therefore, we have $f(1) = f(2) = \cdots = f(3(k+1))$, and by induction f is constant on the positive integers. Now let x be a negative integer. Then $-x$ is a positive integer, and so is $-x + 3$. Thus,

$$f(1) = f\left(\frac{x + (-x + 3)}{3}\right)$$

$$= \frac{f(x) + f(-x + 3)}{2} = \frac{f(x) + f(1)}{2}$$

$$\Rightarrow f(1) = f(x).$$

(15) *Problem.* Determine all functions $f : \mathbf{R} \to \mathbf{R}$ such that

$$f(x - f(y)) = f(f(y)) + xf(y) + f(x) - 1$$

for all numbers x, y.

Solution. Let $a = f(0)$. Now let $x = 0$ in the given relation:

$$f(-f(y)) = f(f(y)) + a - 1 \qquad \text{for all real numbers } y. \qquad (1)$$

If we apply the given relation to $x = f(y)$ we have:

$$a = f(0) = f(f(y)) + [f(y)]^2 + f(f(y)) - 1$$

$$\text{or} \qquad f(f(y)) = \frac{a + 1}{2} - \frac{[f(y)]^2}{2} \qquad (2)$$

again, for all real numbers y. Let us now use (2) to rewrite the given relation:

$$f(x - f(y)) = \frac{a + 1}{2} - \frac{[f(y)]^2}{2} + xf(y) + f(x) - 1 \qquad (3)$$

for all real numbers x and y.

Suppose now that for any real number x we can find real numbers y_1 and y_2 such that $x = f(y_1) - f(y_2)$. Then, by (1) we have

$$f(x - f(y_1)) = f(-f(y_2)) = f(f(y_2)) + a - 1$$

$$= \frac{a + 1}{2} - \frac{[f(y_2)]^2}{2} + a - 1$$

$$= \frac{a + 1}{2} - \frac{[x - f(y_1)]^2}{2} + a - 1$$

$$= \frac{a + 1}{2} - \frac{x^2}{2} + xf(y_1) - \frac{[f(y_1)]^2}{2} + a - 1.$$

But using (3) we also have

$$f(x - f(y_1)) = \frac{a + 1}{2} - \frac{[f(y_1)]^2}{2} + xf(y_1) + f(x) - 1.$$

Comparing the last two equations and cancelling common terms we are left with

$$f(x) = a - \frac{x^2}{2}.$$

Now using (1) with $y = 0$ we have

$$f(-a) = f(a) + a - 1.$$

If $a = 0$, this becomes $a = a + a - 1$, forcing $a = 1$, which is impossible. Therefore, we conclude that $a \neq 0$.
Using the original given relation we have

$$f(y - a) = f(y - f(0)) = f(f(0)) + yf(0) + f(y) - 1$$
$$= f(a) + ya + f(y) - 1$$

for all real numbers y. Then $ya + f(a) - 1 = f(y - a) - f(y)$. Now let x be any real number whatsoever. Since $a \neq 0$, we can set $y = \frac{1}{a}(x + 1 - f(a))$. Clearly,

$$x = ya + f(a) - 1 = f(y - a) - f(y).$$

Thus, every real number x can be expressed as $f(y_1) - f(y_2)$ for some real numbers y_1 and y_2, whence, from above we have $f(x) = a - \frac{1}{2}x^2$.
It remains to determine the value of a. From the above equation we have

$$f(a) = a - \frac{a^2}{2} \quad \text{and} \quad f(-a) = a - \frac{(-a)^2}{2} = f(a),$$

and by applying (1) with $y = 0$, we get

$$f(-a) = f(a) + a - 1,$$

whence $a = 1$. Therefore, the only solution to the problem is the function $f(x) = 1 - \frac{1}{2}x^2$.

(16) *Problem.*   Determine all functions $f : \mathbf{R}\backslash\{0\} \to \mathbf{R}$, satisfying

$$\frac{1}{x}f(-x) + f\left(\frac{1}{x}\right) = x \quad \text{for all } x \in \mathbf{R}\backslash\{0\}.$$

Solution.   Since the given equation is true for any real number x, we can replace x with any other real number and the result will remain true. In particular, we can replace x by $-1/x$ to get

$$-xf\left(\frac{1}{x}\right) + f(-x) = -\frac{1}{x}.$$

Dividing this by x yields:

$$-f\left(\frac{1}{x}\right) + \frac{1}{x}f(-x) = -\frac{1}{x^2}.$$

If we now add this equation to the original (given) equation we obtain:

$$\frac{2}{x}f(-x) = x - \frac{1}{x^2}$$

$$\text{or} \quad f(-x) = \frac{1}{2}\left(x^2 - \frac{1}{x}\right),$$

and if we replace x by $-x$, we get

$$f(x) = \frac{1}{2}\left(x^2 + \frac{1}{x}\right).$$

It is easy to verify that this solution also satisfies the given equation.

(17) *Problem.*   Find all functions $f : (0, \infty) \to (0, \infty)$ such that

$$f(x + \sqrt{x}) \leq x \leq f(x) + \sqrt{f(x)}$$

for all real numbers $x > 0$.

Solution.   Note first that there seems to be an implicit role played by the function $g(x) = x + \sqrt{x}$. Using g, the given inequality can be re-expressed as

$$f\big(g(x)\big) \leq x \leq g\big(f(x)\big).$$

Note that g is also a function defined on the positive real numbers and has the positive real numbers as its range. Furthermore, the function is a one-to-one (monotonically) increasing function. Therefore, g has an inverse function g^{-1}. We will show that the only solution to the given inequality is $f = g^{-1}$.

The right-most part of the given inequality states that, for $x > 0$, we have

$$g\big(f(x)\big) \geq x = g\big(g^{-1}(x)\big).$$

Since g is monotonically increasing, so is g^{-1}. Hence, the above inequality implies that

$$f(x) \geq g^{-1}(x) \tag{1}$$

for all $x > 0$. Now the right-most part of the given inequality states that, for all $x > 0$, we have

$$f\big(g(x)\big) \leq x = g^{-1}\big(g(x)\big).$$

Combining this with equation (1) give us

$$g^{-1}\big(g(x)\big) \leq f\big(g(x)\big) \leq x = g^{-1}\big(g(x)\big)$$

for all $x > 0$. This is an interesting sandwich! It states that we must, in fact, have equality above. That is, we must have $f(g(x)) = g^{-1}(g(x))$ for all $x > 0$. Since $g(x)$ reaches every real number, we conclude that $f = g^{-1}$.

It remains to determine what g^{-1} actually is. Let $y = g^{-1}(x)$. Then $x = g(y)$, which means that $x = y + \sqrt{y}$. Solving this for $\sqrt{y}$ and squaring yields

$$\sqrt{y} = x - y,$$
$$y = (x - y)^2 = x,$$
$$y^2 - y(2x + 1) + x^2 = 0,$$
$$y = \frac{2x + 1 \pm \sqrt{(2x + 1)^2 - 4x^2}}{2},$$
$$= \frac{2x + 1 \pm \sqrt{4x + 1}}{2}.$$

Since $x > y$ (this immediately follows from $x = y + \sqrt{y}$), we must have $y = \left(2x + 1 - \sqrt{4x + 1}\right)/2$.

(To rule out $y = \left(2x + 1 + \sqrt{4x + 1}\right)/2$, consider $x = 1$. Then y would be $(3 + \sqrt{5})/2 > 1$.) That is,

$$f(x) = g^{-1}(x) = \frac{2x + 1 - \sqrt{4x + 1}}{2} = x + \tfrac{1}{2} - \sqrt{x + \tfrac{1}{4}}.$$

(18) *Problem.*   Find all functions $u(x)$ satisfying:

$$u(x) = x + \int_0^{\frac{1}{2}} u(t)dt.$$

Solution.   Since $\int_0^{\frac{1}{2}} u(t)dt$ is a constant, we have $u(x) = x + c$ for some constant c. We need only evaluate c. But then

$$c = \int_0^{\frac{1}{2}} (t + c)dt = \left[\frac{1}{2}t^2 + ct\right]_0^{\frac{1}{2}} = \frac{1}{8} + \frac{1}{2}c,$$

which gives $c = \tfrac{1}{4}$. Thus, $u(x) = x + \tfrac{1}{4}$.

(19) *Problem.*   Suppose that p and q are positive integers, and that f is a function defined for positive real numbers that attains only positive values, such that $f(xf(y)) = x^p y^q$. Prove that $q = p^2$.
Solution.   Let $a = f(1)$. Then

$$f(a) = f(1 \times a) = f(1 \times f(1)) = 1^p \times 1^q = 1.$$

Using this fact, we can show that $a = 1$:

$$1 = f(a) = f(a \times 1) = f(a \times f(a)) = a^p a^q = a^{p+q},$$

and, since p and q are both positive, we conclude that $a = 1$. That is, $f(1) = 1$.

Now let b be any positive number different from 1. Then we have

$$f(b) = f(b \times 1) = f(b \times f(1)) = b^p \times 1^q = b^p.$$

On the other hand, we also have

$$f(b) = f(b^{1-p} \times b^p) = f(b^{1-p} \times f(b)) = (b^{1-p})^p \times b^q = b^{p+q-p^2}.$$

Clearly we have $b^p = b^{p+q-p^2}$, which implies that $p = p + q - p^2$, since $b \neq 1$. But this is obviously equivalent to $q = p^2$.

(20) *Problem.*   Prove that there does not exist a function f mapping the integers into the integers, for which $f(f(x)) = x + 1$ for every integer x.

Solution.   Suppose instead that such a function f does exist. We will derive a contradiction. Apply the function f to both sides of the given equation to get $f(f(f(x))) = f(x+1)$. But the left hand side of this equation can be interpreted as applying the function f twice to the value $f(x)$, since function composition is associative. With this interpretation and applying the given equation, we get $f(f(f(x))) = f(x) + 1$. From this we see that

$$f(x + 1) = f(x) + 1$$

for all integers x. Let $f(O) = a$, where a is some integer. Then

$$f(1) = f(0 + 1) = f(0) + 1 = a + 1$$
$$f(2) = f(1 + 1) = f(1) + 1 = a + 2$$
$$\vdots$$
$$f(n + 1) = f(n) + 1 = a + n + 1.$$

It is clear that $f(k) = a + k$ for all integers k; (this can be formally proved by mathematical induction, which is simply a more formal approach to what we have done above). Thus, on one hand our given equation tells us:

$$f(f(k)) = k + 1,$$

while on the other hand, we have

$$f(f(k)) = f(a + k) = a + (a + k) = 2a + k,$$

since a is an integer. However, this implies that $2a = 1$, which contradicts the fact that a is an integer. Thus, no such function f exists.

(21) *Problem.* Given a continuous function f from the reals into the reals satisfying the conditions:

$$f(1000) = 999,$$

$$f(x) \times f(f(x)) = 1 \qquad \text{for all (real) } x.$$

Determine $f(500)$.

Solution. First note that if $f(x) = a \neq 0$, then

$$f(a) = f(f(x)) = 1/f(x) - 1/a$$

from the second of the two given conditions. Thus, $f(999) = 1/999$, since $f(1000) = 999 \neq 0$. Since f is a continuous function with $f(1000) > 500$ and $f(999) < 500$, there must be a real number y, $999 < y < 1000$ such that $f(y) = 500$. But since $f(y) = 500 \neq 0$, we conclude from the above that $f(500) = 1/500$.

Chapter 3

Higher Dimensional Geometry

(22) *Problem.* The surface area and volume of a certain sphere are both 4-digit integers times π. What is the radius of the sphere?

Solution. Recall the formulas for the volume V and surface area A of a sphere of radius r:

$$V = \frac{4}{3}\pi r^3,$$
$$A = 4\pi r^2.$$

The ratio $V : A = r : 3$. Thus, for $r > 3$ we have $V > A$. Since $r \leq 3$ implies that neither V nor A are in the proper range, we conclude that $r > 3$ and, therefore, $V > A$. Our task is to find an integer value for r such that

$$999\pi < A < V < 10000\pi$$

that is, $\quad 4r^2 > 999$

and $\quad \frac{4}{3}r^3 < 10000.$

Since r is an integer, we see that $16 \leq r \leq 19$. Notice also, however, that $\frac{4}{3}r^3$ must be an integer, whence r must be a multiple of 3. This leaves us with the unique answer of 18 units for the radius.

(23) *Problem.* Euler's formula states that for a convex polyhedron with V vertices, E edges, and F faces, we get $V - E + F = 2$. A particular convex polyhedron has 32 faces (that is, $F = 32$), each of which is either a triangle or a pentagon. At each of its V vertices there are T triangular faces and P pentagonal faces meeting. What are the values of V, E, P and T?

Note: a *polyhedron* is a 3-dimensional solid every face of which is a polygon (for example, triangle, quadrilateral, pentagon, etc.).

A polyhedron is called *convex* if the straight line segment joining any two interior points of the polyhedron lies wholly within the polyhedron.

Solution. Combining $V - E + F = 2$ with $F = 32$ yields $V = E - 30$. Let t be the number of triangular faces on the polyhedron, and let p be the number of pentagonal faces on the polyhedron. Let us now examine the triangular faces; each of the t triangular faces touches 3 vertices and each of the V vertices touches T triangular faces. Therefore, $3t = VT$. Similarly, we can examine the pentagonal faces; each of the p pentagonal faces touches 5 vertices, and each vertex touches P pentagonal faces. Therefore, $5p = VP$. Let us finally examine incidences between vertices and edges; at each of the V vertices we have $P + T$ edges, and on each edge there are 2 vertices. Therefore, $2E = V(P + T) = VP + VT$. If we now recall that all the faces are either triangular or pentagonal, then we see that $32 = F = p + t = \frac{1}{5}VP + \frac{1}{3}VT$, which implies that $180 = 3VP + 5VT$. Solving these last two equations for P and T yields

$$P = \frac{5E - 240}{V} = \frac{5E - 240}{E - 30} = 5 - \frac{90}{E - 30},$$
$$T = \frac{240 - 3E}{V} = \frac{240 - 3E}{E - 30} = \frac{150}{E - 30} - 3.$$

Since both P and T must be non-negative, we see that $48 \leq E \leq 80$. Furthermore, we must have $E - 30$ as a divisor of both 90 and 150, which implies that $E - 30$ must be a divisor of 30. The only possibility for both of these conditions is $E = 60$. In this case we have $V = 30$, $P = 2$ and $T = 2$.

(24) *Problem.* We are given an ordinary cylindrical tin can, eight inches in height and nine inches in circumference, and a long strip of paper of uniform width, two inches to be precise. The strip of paper is wound smoothly around the can without overlapping itself, and neatly trimmed where it crosses the top and bottom rims of the can. If as a result the strip covers exactly one-half of the curved surface of the can, what is the length of one of these cuts where it was trimmed?

Solution. The length of one of these cuts is four and one-half inches, one half the circumference of the can. The height of the can and the width of the strip are totally irrelevant. If the strip covers one-half of the can's curved surface, then it will cover one-half of the surface of any slice of the can parallel to its ends, including in the limit the circumference of the can itself.

(25) *Problem.* The base of a tub is a square with sides of length 1 metre. It contains water 3 centimetres deep. A heavy rectangular block is placed in the tub three times. Each time the face that rests on the bottom of the tub has a different area. When this is done, the water in the tub ends up being 4 centimetres, 5 centimetres and 6 centimetres deep. Find the dimensions of the block.

Solution. Since each of three faces resting on the bottom has a different area, the three dimensions of the block are all different. Therefore, let $x < y < z$ be the lengths of the block in centimetres. Let us now compute the volume of water and block which lie below the level of the water each time the block rests on the bottom:

$$100 \times 100 \times 3 + 4xy = 100 \times 100 \times 4,$$
$$100 \times 100 \times 3 + 5xz = 100 \times 100 \times 5,$$
$$100 \times 100 \times 3 + 6yz = 100 \times 100 \times 6.$$

These yield the relations: $xy = 2500$, $xz = 4000$, and $yz = 5000$. By multiplying these together we get $(xyz)^2 = 5 \times 10^{10}$. Thus, $xyz = \sqrt{5} \times 10^5$. This yields:

$$x = \frac{xyz}{yz} = 20\sqrt{5}, \qquad y = \frac{xyz}{xz} = 25\sqrt{5},$$

$$z = \frac{xyz}{xy} = 40\sqrt{5}.$$

(26) *Problem.* In the diagram of the rectangle-based pyramid below, we have $AB = AC$, $DA = DC$, $\angle BAC = 20°$, and $\angle ADC = 100°$. Prove that $AB = BC + CD$.

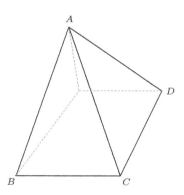

Solution. First note that both triangles ABC and ACD are isosceles, which means that $\angle ABC = \angle ACB = 80°$ and that $\angle ACD = \angle CAD = 40°$. Applying the Law of Sines to $\triangle ABC$ yields:

$$\frac{AC}{\sin 80°} = \frac{BC}{\sin 20°}. \tag{1}$$

Applying it to $\triangle ACD$ yields:

$$\frac{AC}{\sin 100°} = \frac{CD}{\sin 40°}. \tag{2}$$

Solving (1) for BC and (2) for CD and summing produces:

$$
\begin{aligned}
BC + CD &= \frac{AB}{\sin 80°} \cdot (\sin 20° + \sin 40°) \\
&= \frac{AB}{\cos 10°} \cdot \left(\sin(30° - 10°)\sin(30° + 10°) \right) \\
&= \frac{AB}{\cos 10°} \cdot (\sin 30° \cos 10° - \cos 30° \sin 10° \\
&\quad + \sin 30° \sin 10° + \cos 30° \sin 10°) \\
&= \frac{AB}{\cos 10°} \cdot (2 \sin 30° \cos 10°) \\
&= AB \qquad \text{since } \sin 30° = \frac{1}{2}.
\end{aligned}
$$

Note that the problem has the same solution if the triangles ABC and ACD simply lie in the plane.

The three-dimensional pyramid is simply a red herring!

(27) *Problem.* A conical drinking glass is 12 cm deep and has a top diameter, of 10 cm. The glass contains some liquid. A sphere of diameter 8 cm is placed in the glass, and it turns out that the liquid exactly fills the volume below the points where the sphere touches the cone.

Find the depth of the lowest point of the sphere beneath the highest point of the liquid.

Solution. Let us consider a vertical cross section through the centre of the conical drinking glass, as in the diagram on the next page:

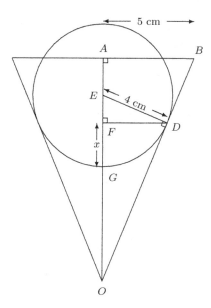

Let the vertex of the cone be O; let A be the centre of the top; and let B be a point on the top edge of the cone. Let D be the point of contact between the sphere and the cone on the side OB; let E be the centre of the sphere; let G be the lowest point on the sphere; and let F be the point on EG which is at the same height above O as is D.

Further, let x be the length of the line segment FG in cm. We must find x. Triangle OAB is right-angled, and by the Theorem of Pythagoras, its hypotenuse, OB, is 13 cm in length. Clearly, triangles OAB and ODE are both right-angled, and thus they are similar. But triangle DFE is also similar to ODE. Therefore, triangles OAB and DFE are similar. This means that

$$\frac{EF}{DE} = \frac{AB}{OB},$$

that is, $\quad EF = DE \cdot \dfrac{AB}{OB} = 4 \cdot \dfrac{5}{13} = \dfrac{20}{13}.$

Then, $x = EG - EF = 4 - (20/13) = 32/13$ cm.

(28) *Problem.* A convex polyhedron is a 3-dimensional solid whose surface is composed on polygons, such that the line segment joining any two points of the solid lies wholly within the solid (that is, it does not "fold in" on itself). Determine all convex polyhedra (plural of polyhedron) satisfying:

(a) every face is either an equilateral triangle or a square;

(b) every edge separates a square from an equilateral triangle.

Hint: you may find it useful to employ Euler's Formula for convex polyhedra, namely $V+F-E = 2$, where V is the number of vertices (or "corners"), F is the number of faces (or polygonal surfaces), and E is the number of edges.

Solution. Let V, F, and E be defined as in the hint above. Consider a single vertex of the polyhedron, and examine the faces that meet at this vertex. Since the edges emanating from the vertex must separate squares and equilateral triangles, the faces must alternate in order around the vertex, which implies there is an even number of faces meeting at the vertex (the same number of squares as triangles). This even number cannot be as small as 2 since any vertex of any polyhedron must have at least 3 faces meeting at it. If we add up the interior angle measures of the faces meeting at the given vertex, we must have a value smaller than $360°$, in order that the faces not all lie in one plane; this implies that we have fewer than 6 faces since each square-triangle pair accounts for $90° + 60° = 150°$ and 3 such pairs would be too many. Thus, each vertex of the polyhedron must lie on exactly 2 squares and 2 equilateral triangles, which alternate around the vertex.

Now let S be the number of square faces, and let T be the number of triangular faces, on the polyhedron. If we examine each triangular face and count the number of edges which bound the face we get the value $3T$; but this also counts each edge exactly once since every edge separates a square and a triangle. Thus, $E = 3T$. A similar calculation with the square faces shows that $E = 4S$.

Now let us examine the vertices. If we look at the triangular faces and count vertices we get $3T$ again. However, in this case we have counted each vertex twice since each vertex lies on exactly two triangular faces. Thus, we have $2V = 3T$. From this we can conclude that $E = 2V = 3T = 4S$. Since $E = 2V$, we see that Euler's Formula can be written as $F = V + 2$.

Since $F = S+T$ (as there are only the two types of faces), we have $V+2 = S+T$, or $4V+8 = 4S+4T$, and substituting $4V = 3T+4S$, we conclude that $T = 8$. This means that $E = 2V = 4S = 24$, whence, $S = 6$, $V = 12$, $E = 24$ and $F = 12 + 2 = 14$.

In order to understand that this describes a unique polyhedron, we will focus on one of the square faces. There is a triangle attached to each of its 4 faces, and lying between each pair of successive triangles going around the square lies another square. This accounts for 5 of the 6 squares. Between each successive pair of squares that we have just introduced we have another triangle, giving us a total of 8 triangles (which is all of them), which means we need one last square to close up the solid. This means that if such a polyhedron exists, there is only one such.

Now we ask: Does such a polyhedron actually exist? The answer is YES. Consider a cube. Mark the mid-point of each edge, and then cut from each corner a small tetrahedron using a planar cut through the marks on the three edges emanating from the vertex. This gives a convex polyhedron satisfying all the conditions of the problem.

(29) *Problem.* Two perpendicular planes intersect a sphere in two circles. These circles intersect in two points spaced 14 cm apart, measured along the straight line connecting them. If the radii of the two circles are 18 cm and 25 cm, what is the radius of the sphere?

Solution. Let O be the centre of the sphere, and let r be its radius. Let C_1 and C_2 be the centres of the circles of radius 18 cm and 25 cm, respectively. Let A and B be the points of intersection of the two circles. Thus, AB is a chord of both circles and also of the sphere. Let D be the mid-point of the chord. We need to determine $r = OA = OB$. Consider first the circle centred at C_1. Triangle C_1AD is a right-angled triangle with right angle at D. By the Theorem of Pythagoras we have:

$$C_1D^2 = C_1A^2 - AD^2 = 18^2 - 7^2 = 275.$$

Similarly, by considering $\triangle C_2AD$ in the circle centred at C_2, we obtain:

$$C_2D^2 = C_2A^2 - AD^2 = 25^2 - 7^2 = 576.$$

Let us now compute $C_1C_2^2$. By the Theorem of Pythagoras (again), we have:

$$C_1C_2^2 = C_1D^2 + C_2D^2 = 275 + 576 = 851.$$

Since the planes of the two circles are perpendicular, and since lines perpendicular to each circle passing through its centre also pass through the centre of the sphere, we see that the points C_1, D, C_2, and O are not only coplanar, but also form a rectangle. Therefore, we have $OD = C_1C_2$. Now $\triangle OAD$ is also a right-angled triangle, so that, applying the Theorem of Pythagoras again, we have:

$$r^2 = OA^2 = AD^2 + OD^2 = 7^2 + C_1C_2^2 = 49 + 851 = 900.$$

Therefore, $r = 30$.

(30) *Problem.* A student has two open-topped containers: the larger one is a right circular cylinder of height 21 cm and radius 12 cm, the smaller one is a right circular cone of height 20 cm and radius 10 cm (at the open end). (The walls of the two containers are thin enough so that their widths can be ignored.) The larger container initially has water in it to a depth of 18 cm. The student slowly lowers the cone into the larger container, keeping the top (open side) of the cone horizontal. As the cone is lowered, water may first overflow out of the larger container, but eventually water pours into the cone. When the tip of the cone is resting on the bottom of the cylinder, what is the depth of the water in the cone?

Solution. Let us first compute the volume of water in the larger container at the beginning of the process. That volume is $V_0 = \pi(12)^2(18) = 2592\pi$ cm^3. Now let us push the cone into the cylinder and compute the total volume inside the cylinder and outside the cone. This is the maximum possible amount of water we are dealing with, since any water in excess of this amount will have overflowed the cylinder, and is gone. This volume is

$$V_w = \pi(12)^2(21) - \frac{\pi(10)^2(20)}{3} = 3024\pi - \frac{2000\pi}{3} = \frac{7072\pi}{3} \text{ cm}^3.$$

Since this is less than V_0, we see that V_w is the amount of water we are dealing with. Now let us push the cone down in the cylinder until its tip touches the bottom of the cylinder. The volume of water outside the cone and inside the cylinder but below the top of the cone is seen to be:

$$V_{\text{out}} = \pi(12)^2(20) - \frac{(10)^2(20)}{3} = 20\pi\left(144 - \frac{100}{3}\right) = \frac{6640\pi}{3} \text{ cm}^3.$$

The remainder of the water, that is, $V_w - V_{\text{out}}$ must have flowed over the top of the cone and is now in the cone. This volume is

$$V_{\text{in}} = \frac{7072\pi}{3} - \frac{6640\pi}{3} = \frac{432\pi}{3} = 144\pi \text{ cm}^3.$$

The water in the cone has taken on the shape of a cone similar to the one containing it. Denote by r and h the radius and height of the water inside the cone. We first of all note that $r : h = 10 : 20$, or $r = h/2$. Thus, we have

$$144\pi = V_{\text{in}} = \frac{1}{3}\pi r^2 h = \frac{h^3 \pi}{12},$$

$$\therefore \quad h^3 = 12(144) = 12^3,$$

$$h = 12.$$

Therefore, the depth of water in the cone is 12 cm.

The interested reader might be interested in determining how the answer changes with different initial amounts of water in the cylinder. For example, what is the answer if the initial depth was 15 cm? Or 10 cm? There are two critical values for the initial depth of water. What are they?

(31) *Problem.* Let P be a convex polyhedron in which four edges meet at every vertex. Prove that if we cut P with any plane which does not pass through a vertex, the resulting intersection is a polygon with an even number of sides. (A set of points is said to *convex* if the line segment joining any two points of the set contains only points from the set.)

Solution. Let P be the polyhedron in the problem and let π be the plane cutting P and not passing through a vertex of P. Then π cuts P into two convex polyhedra with a common face F. Indeed, F is the polygon in the problem statement. Let m be the number of sides of F. We must show that m is even.

Let P_1 be one of the two convex polyhedra resulting from the cut. Then P_1 has m vertices of degree 3 (that is, there are m vertices which have 3 edges incident with them), namely the m vertices of F. Its remaining vertices, say n in number, still have degree 4. The sum of the degrees of the vertices of P_1 is thus $3m + 4n$. Since each edge is incident with two vertices, the number of edges of P_1, which must be an integer, is

$$\frac{3m + 4n}{2} = \frac{3m}{2} + 2n,$$

which means that m must be even.

For your further information, Peter Smoczynski has generalized this problem. In order to fully understand his solution, we must introduce some concepts from Graph Theory.

A *graph* G is a set V of objects called *vertices* together with a set E of unordered pairs of vertices called *edges*. We denote this by writing $G = (V, E)$. An edge (v_1, v_2) is said to be *incident* with both of the vertices v_1 and v_2. The number of edges incident with a vertex v is called the *degree* of the vertex, and is denoted by $\deg(v)$. Note that the vertices and edges of a polyhedron can be treated as the vertices and edges of a graph. A *subgraph* of a graph (V, E) is a graph whose vertices are a subset of V.

Definition: Let S be a subgraph of a graph G. Then an edge of G is called an *external* edge of S if one of its vertices belongs to S and the other does not.

Proposition: Let G be a finite graph (that is, it has only finitely many vertices), and let every vertex of G have even degree. If S is any subgraph of G, then the number of external edges of S is even.

Proof: We first observe that the subgraph S can be obtained from the graph G by systematically removing vertices (and the necessary edges), and finally depleting the edge set down to that of S. In this way, we construct a sequence of subgraphs:

$$G = S_0 \supseteq S_1 \supseteq \cdots \supseteq S_n = S.$$

We will prove the result by induction on the number n of vertices that need to be removed from G to obtain S. If $n = 0$, then $S = G$ and the number of external edges is 0, which is even. Now suppose the result is true for $n = k$, for some $k \geq 0$; that is, S_k has an even number of external edges. Let v be the next vertex to be removed from S_k to obtain S_{k+1}, and let m be the number of edges joining v to the other vertices of S_{k+1}. After removing v to get S_{k+1}, we have lost $\deg(v) - m$ external edges (they were external in S_k), and we have gained m new external edges. Thus, the number of external edges has increased by $m - \big(\deg(v) - m\big) = 2m - \deg(v)$, which is an even number since $\deg(v)$ is even. Therefore, S_{k+1} also has an even number of external edges, and the result is true for $n = k + 1$. By induction, the result is true for all n, and the proof is complete.

Application: The original problem could have been posed as: Let P be a convex polyhedron in which an even number of edges meet at every vertex. Prove that if we cut the polyhedron into pieces such that the cut does not pass through any vertex, then the cut crosses an even number of edges of the polyhedron.

(32) *Problem.* The surface area of a closed cylinder is twice the volume. Determine the radius and height of the cylinder given that the radius and height are both integers.

Solution. Let r be the radius and let h be the height of the closed cylinder, where r and h are both positive integers. If the surface of the closed cylinder is twice the volume, then

$$2\pi rh + 2\pi r^2 = 2(\pi r^2 h).$$

It follows that $h + r = rh$. Thus,

$$(r-1)(h-1) = rh - r - h + 1 = 0 + 1 = 1.$$

Hence, $r - 1 = h - 1 = \pm 1$ (since r and h are both integers), so that $r = h = 2$ or $r = h = 0$. Since the cylinder actually exists, we will eliminate the solution $r = h = 0$, leaving us with $r = h = 2$ as the solution.

It can be checked that if $r = 2$ and $h = 2$, then the surface area of the closed cylinder is twice the volume.

(33) *Problem.* A number of spheres, all with radius 1, are being placed in the form of a square pyramid. First, there is a layer in the form of a square with $n \times n$ spheres. On top of that layer comes the next layer with $(n-1) \times (n-1)$ spheres, and so on. The top layer consists of only one sphere. Determine the height of the pyramid.

Solution. The centres of the spheres on the lowest layer are 1 unit off the "ground". The top of the top sphere is 1 unit above its centre. The centres of each layer lie a fixed height, say h above the centres of the next lower layer. Thus, the total height of the pyramid is $2 + (n-1)h$. We need only determine the value of h.

To calculate the value of h, we need only consider a layer of 4 unit spheres (laid out in a square see figure (a) below) with a single sphere on top of it.

(a)

The centres of any two touching spheres are 2 units apart. Consider the plane containing the centres of the four spheres on the bottom. The distance between the centres of each pair of non-touching spheres in that layer is then clearly $2\sqrt{2}$.

If we now consider a vertical plane through the centre of the top sphere and a pair of non-touching spheres in the next lower layer (see figure (b) below), we see that the altitude is h and this is clearly $\sqrt{2}$. Thus, the total height of the pyramid is $2 + (n-1)\sqrt{2}$.

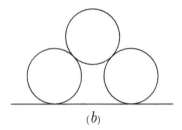

(b)

Chapter 4

Identities, Inequalities and Expressions

(34) *Problem.* Without calculating the value, prove that

$$\frac{1}{2} \cdot \frac{3}{4} \cdot \frac{5}{6} \cdots \cdot \frac{99}{100} < \frac{1}{10}.$$

Solution. Note that

$$\frac{1}{2} < \frac{2}{3}$$

$$\frac{3}{4} < \frac{4}{5}$$

$$\vdots$$

$$\frac{2n-1}{2n} < \frac{2n}{2n+1}.$$

Therefore,

$$A = \frac{1}{2} \cdot \frac{3}{4} \cdot \frac{5}{6} \cdots \cdot \frac{99}{100} < \frac{2}{3} \cdot \frac{4}{5} \cdot \frac{6}{7} \cdots \cdot \frac{100}{101} = B.$$

On the other hand, we have

$$AB = \frac{1}{2} \cdot \frac{2}{3} \cdot \frac{3}{4} \cdot \frac{4}{5} \cdots \cdot \frac{99}{100} \cdot \frac{100}{101} = \frac{1}{101}.$$

However, we also have $A^2 < AB$, which means that

$$A < \sqrt{AB} = \frac{1}{\sqrt{101}} < \frac{1}{10}.$$

(35) *Problem.* Prove or disprove: there are real numbers x and y such that
$$\begin{aligned} x &+ y &= 1, \\ x^2 &+ y^2 &= 2, \\ x^3 &+ y^3 &= 3. \end{aligned}$$

37

Solution. There are <u>no</u> real x and y satisfying the above system.

Proof: For the purpose of contradiction, we suppose that there are real x and y that satisfy the above system. Then

$$(x+y)^2 = 1^2, \qquad \text{that is;} \qquad x^2 + 2xy + y^2 = 1.$$

Since $x^2 + y^2 = 2$, we see that $2xy = -1$, or $xy = -\frac{1}{2}$. Expanding $(x+y)^3 = 1^3$ and using the third equation above we obtain that

$$x^3 + 3x^2 y + 3xy^2 + y^3 = 1,$$
$$3x^2 y + 3xy^2 = -2,$$
$$3xy(x+y) = -2,$$
$$3 \cdot \left(-\frac{1}{2}\right) \cdot 1 = -2$$
$$-\frac{3}{2} = -2,$$

which is absurd. Thus, our supposition is incorrect, and we conclude that there are no real x and y that satisfy the given system.

(36) *Problem.* Evaluate the infinite product:

$$\left(\frac{1}{2}\right)^{\frac{1}{3}} \cdot \left(\frac{1}{4}\right)^{\frac{1}{9}} \cdot \left(\frac{1}{8}\right)^{\frac{1}{27}} \cdot \left(\frac{1}{16}\right)^{\frac{1}{81}} \cdots \cdots \left(\frac{1}{2^n}\right)^{\frac{1}{3^n}} \cdots$$

Solution. Let P be the value of the infinite product. Then

$$P = \left(\frac{1}{2}\right)^{\frac{1}{3}} \cdot \left(\frac{1}{2}\right)^{\frac{2}{9}} \cdot \left(\frac{1}{2}\right)^{\frac{3}{27}} \cdot \left(\frac{1}{2}\right)^{\frac{4}{81}} \cdots \cdots \left(\frac{1}{2}\right)^{\frac{n}{3^n}} \cdots \cdots = \left(\frac{1}{2}\right)^{S},$$

where

$$S = \frac{1}{3} + \frac{2}{9} + \frac{3}{27} + \frac{4}{81} + \cdots + \frac{n}{3^n} + \cdots$$

We will now rearrange the order of summation (which can be done since all the terms are positive):

$$S_1 = \frac{1}{3} + \frac{1}{9} + \frac{1}{27} + \frac{1}{81} + \cdots + \frac{1}{3^n} + \cdots$$
$$S_2 = \qquad + \frac{1}{9} + \frac{1}{27} + \frac{1}{81} + \cdots + \frac{1}{3^n} + \cdots$$
$$S_3 = \qquad \qquad + \frac{1}{27} + \frac{1}{81} + \cdots + \frac{1}{3^n} + \cdots$$
$$\vdots$$

Clearly, $S = S_1 + S_2 + S_3 + \cdots + S_n + \cdots$. But S_i is a geometric series with common ratio $r = \frac{1}{3}$ for each $i \geq 1$. Therefore, $S_i = a_i/(1-r)$, where a_i is the first term of the series; that is, $a_i = 1/3^i$. Thus,

$$S_i = \frac{1/3^i}{1 - \frac{1}{3}} = \frac{1/3^i}{\frac{2}{3}} = \frac{3}{2} \cdot \frac{1}{3^i} = \frac{1}{2} \cdot \frac{1}{3^{i-1}}.$$

Consequently,

$$S = \frac{1}{2}\left(1 + \frac{1}{3} + \frac{1}{9} + \frac{1}{27} + \cdots + \frac{1}{3^{n-1}} + \cdots\right),$$

which is another geometric series with common ratio $r = \frac{1}{3}$. Therefore,

$$S = \frac{1}{2}\left(\frac{1}{1 - \frac{1}{3}}\right) = \frac{1}{2} \cdot \frac{3}{2} = \frac{3}{4};$$

whence

$$P = \left(\frac{1}{2}\right)^S = \left(\frac{1}{2}\right)^{\frac{3}{4}}.$$

(37) *Problem.* Solve $\sqrt{x + 20} - \sqrt{x + 1} = 1$.
If you can do that, then solve $\sqrt[3]{x + 20} - \sqrt[3]{x + 1} = 1$.
Solution. Set $A = \sqrt{x + 20}$ and $B = \sqrt{x + 1}$. Then we have

$$A^2 - B^2 = (A + B)(A - B) = A + B,$$

since $A - B = 1$. Thus, $x + 20 - (x + 1) = 19 = A + B$, implying that $A = 10$ and $B = 9$, from which we see that $x = 80$.
Let $A = \sqrt[3]{x + 20}$ and $B = \sqrt[3]{x + 1}$. Thus, $A - B = 1$.
Multiplying by $A^2 + AB + B^2$ yields $A^3 - B^3 = A^2 + AB + B^2$, which means that $A^2 + AB + B^2 = x + 20 - (x + 1) = 19$.
Also, we see that $(A - B)^2 = A^2 + AB + B^2 = 1$. Subtracting, we get $3AB = 19 - 1 = 18$, or $AB = 6$.
This implies that $(x + 20)(x + 1) = A^3 B^3 = 6^3 = 216$ which yields the quadratic equation $x^2 + 21x - 196 = 0$.
This is equivalent to $(x + 28)(x - 7) = 0$.
Therefore, the solution is $x = -28$ or $x = 7$.

(38) *Problem.* Find the product of x, y, and z if $\{x, y, z\}$ is the solution set for the given equations:

$$\begin{aligned} x + y + z &= 1, \\ x^2 + y^2 + z^2 &= 2, \\ x^3 + y^3 + z^3 &= 3. \end{aligned}$$

Solution. First observe that

$$(x + y + z)^3$$
$$= x^3 + y^3 + z^3 + 3x^2y + 3x^2z + 3xy^2 + 3xz^2 + 3y^2z$$
$$+ 3yz^2 + 6xyz$$
$$= (x^3 + y^3 + z^3) + 3x^2(y + z) + 3y^2(x + z) + 3z^2(x + y) + 6xyz$$
$$= (x^3 + y^3 + z^3) + 3(x^2 + y^2 + z^2)(x + y + z)$$
$$- 3x^3 - 3y^3 - 3z^3 + 6xyz$$
$$= 3(x^2 + y^2 + z^2)(x + y + z) - 2(x^3 + y^3 + z^3) + 6xyz.$$

By using the given relationships this equation becomes

$$1 = 3(2)(1) - 2(3) + 6xyz,$$

from which we deduce that $xyz = \frac{1}{6}$.

(39) *Problem.*   If

$$2\ln(x - 2y) = \ln x + \ln y,$$

find x/y.

Solution. The first step is to rewrite the equation as

$$\ln(x - 2y)^2 = \ln(xy),$$

from which one can conclude that

$$(x - 2y)^2 = xy,$$
$$\text{or} \quad x^2 - 4xy + 4y^2 = xy.$$

This can be rearranged to

$$x^2 - 5xy + 4y^2 = 0,$$

which can be solved (using the quadratic formula) to yield $x = y$ or $x = 4y$. Since logarithms can only be evaluated for positive values of the argument, we can eliminate $x = y$. Thus, $x = 4y$ or rather $x/y = 4$.

(40) *Problem.*   Prove that

$$\frac{1}{\log_2 \pi} + \frac{1}{\log_5 \pi} > 2.$$

Solution. First of all note that for logarithms to any bases a and b we have

$$\log_a b = \frac{1}{\log_b a}.$$

(This can be established by setting $x = \log_a b$ and recognizing that this is equivalent to $a^x = b$. From this by taking logarithms to the base b on both sides, we have $x \log_b a = 1$; whence, $\log_a b = x = 1/\log_b a$.)

With this preamble we can then rewrite the statement to be proved as

$$\log_\pi 2 + \log_\pi 5 > 2.$$

The left hand side can be rewritten as $\log_\pi 10$, and the question we must answer is whether $10 > \pi^2$. Since this is, in fact, true, we are finished.

(41) *Problem.* Given that a and b are two positive real numbers, show that if

$$a^a = b \qquad \text{and} \qquad b^b = a,$$

then $a = b = 1$.

Solution. We will prove this result by contradiction. Therefore, let us assume that at least one of a and b are different from 1, say $a \neq 1$. By raising both sides of the first condition above to the power b, we get

$$a^{ab} = b^b = a,$$

the last part following from the second condition above. We see then that a raised to the power ab gives the same value as a raised to the power 1. Thus, we may conclude that $ab = 1$, since a itself is not 1. By using our first condition again, we get

$$a^0 = 1 = ab = a \cdot a^a = a^{a+1},$$

from which we may conclude by arguments similar to the above that $a = -1$, which is impossible. This final contradiction establishes the result.

(42) *Problem.* Let

$$A_n = \left(7 + 4\sqrt{3}\right)^n,$$

where n is a positive integer. Find a simple expression for $1 + [A_n] - A_n$, where $[x]$ is the greatest integer less than or equal to x.

Solution. First note that since n is a positive integer, we observe that $A_n > 1$. Next define $B_n = \left(7 - 4\sqrt{3}\right)^n$. Then

$$A_n B_n = \left(7 + 4\sqrt{3}\right)^n \left(7 - 4\sqrt{3}\right)^n$$

$$= \left[\left(7 + 4\sqrt{3}\right)\left(7 - 4\sqrt{3}\right)\right]^n$$
$$= (49 - 48)^n = 1^n = 1.$$

Thus, we may conclude that $B_n = 1/A_n$, whence $0 < B_n < 1$. Since A_n is clearly not an integer, the expression $1 + [A_n]$ is the smallest integer not smaller than A_n. On the other hand $A_n + B_n$ can be seen to be an integer by expanding each part using the binomial theorem (where all terms involving the irrational value $\sqrt{3}$ cancel out). Since $B_n < 1$ and $A_n + B_n$ is an integer, it must obviously be the next higher integer; that is, it must be the smallest integer not smaller than A_n. Therefore,

$$1 + [A_n] - A_n = A_n + B_n - A_n = B_n = \left(7 - 4\sqrt{3}\right)^n.$$

Thus, the simple expression that we seek is $\left(7 - 4\sqrt{3}\right)^n$ or $1/A_n$.

(43) *Problem.*   Find the positive solution of

$$\frac{1}{x^2 - 3x - 6} + \frac{2}{x^2 - 3x - 15} - \frac{3}{x^2 - 3x - 30} = 0.$$

Solution.   If we bring the left hand side to a common denominator and set the resulting numerator to 0, we get

$$(x^2 - 3x - 15)(x^2 - 3x - 30) + 2(x^2 - 3x - 6)(x^2 - 3x - 30)$$
$$- 3(x^2 - 3x - 6)(x^2 - 3x - 15) = 0.$$

The terms involving $(x^2 - 3x)(x^2 - 3x)$ clearly cancel. There only remains

$$(-15 - 30)(x^2 - 3x) + 15(30) + 2(-6 - 30)(x^2 - 3x) + 2(-6)(-30)$$
$$- (-6 - 15)(x^2 - 3x) - 3(-6)(-15) = 0,$$

which simplifies to

$$-54(x^2 - 3x) + 540 = 0$$

or

$$x^2 - 3x - 10 = 0.$$

This equation has solutions $x = -2$ and $x = 5$. Thus, the only positive solution of the original equation is $x = 5$, which can readily be checked to be a solution.

(44) *Problem.*   Determine the value of $w + x + y + z$ given that

$$6w + 2x = 1563,$$
$$2x + 3y + 6z = 4392,$$
$$y - 2w = 1476.$$

Solution. Notice that

$$6(w + x + y + z) = 2(6w + 2x) + (2x + 3y + 6z) + 3(y - 2w)$$
$$= 2(1563) + 4392 + 3(1476) = 11946.$$

Thus, $w + x + y + z = 1991$.

(45) *Problem.* Solve for x:

$$(x^2 + 3x + 2)(x^2 + 7x + 12) + (x^2 + 5x - 6) = 0.$$

Solution. First of all, notice that all of the quadratics in the equation can be factored. Let us factor the first two of them and then recombine the factors:

$$(x + 1)(x + 2)(x + 3)(x + 4) + (x^2 + 5x - 6) = 0,$$
$$(x + 1)(x + 4)(x + 2)(x + 3) + (x^2 + 5x - 6) = 0,$$
or $$(x^2 + 5x + 4)(x^2 + 5x + 6) + (x^2 + 5x - 6) = 0.$$

Now replace $x^2 + 5x$ by y. The equation becomes

$$(y + 4)(y + 6) + (y - 6) = 0,$$

which simplifies to

$$y^2 + 11y + 18 = 0,$$

that is, $$(y + 2)(y + 9) = 0.$$

Therefore, upon substituting $x^2 + 5x$ for y, we get

$$x^2 + 5x + 2 = 0 \quad \text{or} \quad x^2 + 5x + 9 = 0.$$

Using the quadratic formula on these two equations we get the solutions for x:

$$x = \frac{-5 \pm \sqrt{21}}{2} \quad \text{or} \quad \frac{-5 \pm i\sqrt{11}}{2}.$$

(46) *Problem.* If a, b, and c are odd integers, prove that any real roots of

$$ax^2 + bx + c = 0$$

are irrational.

Solution. Let us assume that there is a rational (real) root of the above quadratic where all three coefficients are odd integers. We must then derive a contradiction. Let p/q be such a rational root, where p and q are both integers and $q \neq 0$. Let us further assume that p and q have no common integer factors. Then

$$a\left(\frac{p}{q}\right)^2 + b \cdot \frac{p}{q} + c = 0,$$

that is, $ap^2 + bpq + cq^2 = 0.$

Since p and q have no common integer factor, either both are odd or exactly one of them is even. In any event the expression $ap^2 + bpq + cq^2$ is odd, since a, b and c are all odd. But this expression must equal 0, which is not odd! Thus, we have a contradiction and the proof is complete.

(47) *Problem.*   Find all positive values of the parameter a for which the common solutions of the inequalities

$$x^2 - 2x \leq a^2 - 1 \qquad \text{and} \qquad x^2 - 4x \leq -a - 2$$

form an interval of length 1 on the real axis.

Solution.   The two inequalities are equivalent to

$$(x - 1)^2 \leq a^2 \qquad \text{and} \qquad (x - 2)^2 \leq 2 - a.$$

The first has as its solution the closed interval from $1 - a$ to $1 + a$. The second has as its solution the closed interval from $2 - |2 - a|$ to $2 + |2 - a|$, which is from a to $4 - a$ if $a \leq 2$ and from $4 - a$ to a if $a > 2$. Since each interval must itself have length at least 1, we conclude from the first interval that $a \geq \frac{1}{2}$ and from the second interval that $a \leq \frac{3}{2}$ or $a \geq \frac{5}{2}$, from which we get $\frac{1}{2} \leq a \leq \frac{3}{2}$ or $a \geq \frac{5}{2}$. Let us then distinguish two cases:

(i) $\frac{1}{2} \leq a \leq \frac{3}{2}$
The common solutions must lie between a and $1 + a$, which is an interval of width 1, for all a. Thus one solution is the entire closed interval $[\frac{1}{2}, \frac{3}{2}]$.

(ii) $a \geq \frac{5}{2}$
The common solutions must lie between $4 - a$ and $1 + a$, which is an interval of width $2a - 3$. The width interval takes on its smallest value when a is its smallest value, namely $a = \frac{5}{2}$, which yields a width of 2. Thus, we have no solution for $a \geq \frac{5}{2}$.

Therefore, the only solution to the problem is for a to be anywhere in the interval $[\frac{1}{2}, \frac{3}{2}]$.

(48) *Problem.* Find all positive integer values of a and b that satisfy the equation

$$\frac{1}{a} + \frac{a}{b} + \frac{1}{ab} = 1.$$

Solution. If we multiply both sides by ab, we get

$$b + a^2 + 1 = ab,$$

which can be written as

$$ab - b = a^2 + 1,$$

from which we get either $a = 1$, which is impossible, or

$$b = \frac{a^2 + 1}{a - 1} = a + 1 + \frac{2}{a - 1}. \qquad (*)$$

Since b is an integer we must conclude that $a - 1$ divides evenly into 2, which implies that $a = 2$ or $a = 3$. For either value of a, when we compute the value of b from $(*)$, we have $b = 5$. Thus the only solution in integers is $(a, b) = (2, 5)$ or $(3, 5)$.

(49) *Problem.* Find all real solutions for the following equation:

$$\sqrt[3]{2x - 7} + \sqrt[3]{3x - 3} = \sqrt[3]{x - 8} + \sqrt[3]{4x - 2}.$$

Solution. In order to solve this equation we first prove a short lemma:

Lemma. α, β, γ, and δ satisfy the system $\alpha + \beta = \gamma + \delta$ and $\alpha^3 + \beta^3 = \gamma^3 + \delta^3$ if and only if
(i) $\alpha = -\beta$ and $\gamma = -\delta$, or
(ii) $\{\alpha, \beta\} = \{\gamma, \delta\}$.
Proof: If α, β, γ, and δ satisfy the system, then

$$\begin{aligned}
0 &= (\alpha + \beta)^3 - (\gamma + \delta)^3 \\
&= \alpha^3 + \beta^3 - \gamma^3 - \delta^3 + 3\alpha\beta(\alpha + \beta) + 3\gamma\delta(\gamma + \delta) \\
&= 3\alpha\beta(\alpha + \beta) + 3\gamma\delta(\gamma + \delta) = 3(\alpha\beta - \gamma\delta)(\alpha + \beta).
\end{aligned}$$

If $\alpha + \beta = 0$, then we have case (i).
If $\alpha\beta - \gamma\delta = 0$ and $\alpha + \beta = \gamma + \delta$, then $\{\alpha, \beta\} = \{\gamma, \delta\}$ since both sets are the roots of the same quadratic equation, namely:

$$x^2 - (\alpha + \beta)x + \alpha\beta = 0.$$

This gives (ii). To prove the other direction is trivial.

Applying the lemma to $\alpha = \sqrt[3]{2x-7}$, $\beta = \sqrt[3]{3x-3}$, $\gamma = \sqrt[3]{x-8}$, and $\delta = \sqrt[3]{4x-2}$, gives $\alpha = -\beta$ and $\gamma = -\delta$ in which case $x = 2$, or $\{\alpha, \beta\} = \{\gamma, \delta\}$. Setting $\alpha = \gamma$ and $\beta = \delta$ we get $x = -1$; setting $\alpha = \delta$ and $\beta = \gamma$ we get $x = -\frac{5}{2}$.

(50) *Problem.* Determine all numbers a for which the equation

$$a3^x + 3^{-x} = 3$$

has a unique real solution x.

Solution. Let us first set $y = 3^x$, and observe that $y > 0$ for all real x. Our equation then becomes

$$ay + \frac{1}{y} = 3,$$

which can be rearranged as

$$ay^2 - 3y + 1 = 0.$$

If $a = 0$, we get the (unique) solution $y = 1/3$, which corresponds to $x = -1$. If $a \neq 0$, then we have a quadratic equation. This quadratic equation has roots given by

$$y = \frac{3 \pm \sqrt{9 - 4a}}{2a}.$$

We are looking for values of a which yield exactly one solution for x, which means there should be exactly one positive value y for such a value a. This clearly happens when the discriminant above, namely $9 - 4a$, is 0, that is, when $a = 9/4 = 2.25$. However, it also occurs when one of the values for y is negative, since we know that $y > 0$ for all real x. This happens when $\sqrt{9 - 4a} > 3$, that is, when $a < 0$. Thus, the values of a which yield a unique real solution x are $a \leq 0$ or $a = 2.25$.

(51) *Problem.* Find all real solutions to the equation

$$\sqrt{3x^2 - 18x + 52} + \sqrt{2x^2 - 12x + 162} = \sqrt{-x^2 + 6x + 280}.$$

Solution. Notice that, by completing the square, we have

$$3x^2 - 18x + 52 = 3(x-3)^2 + 25,$$
$$2x^2 - 12x + 162 = 2(x-3)^2 + 144,$$
$$-x^2 + 6x + 280 = -(x-3)^2 + 289.$$

Therefore, the L.H.S. is at least as large as $\sqrt{25} + \sqrt{144} = 5 + 12 = 17$ and the R.H.S. is at most $\sqrt{289} = 17$, with equality for both when $x = 3$, Therefore, $x = 3$ is the only solution.

(52) *Problem.* Find all non-negative numbers x, y and z such that

$$z^x = y^{2x},$$
$$2^z = 2 \cdot 4^x,$$
$$x + y + z = 16.$$

Solution. From the first equation we see that $z = y^2$, or else $z = 0$ and $y = 0$. In either case we have $z = y^2$.
The second equation can be rewritten as $2^z = 2^{2x+1}$, implying that $z = 2x + 1$.
Thus, $x = \frac{1}{2}(z - 1) = \frac{1}{2}(y^2 - 1)$. Placing these expressions for z and x into the third equation we have:

$$\tfrac{1}{2}(y^2 - 1) + y + y^2 = 16,$$
$$y^2 - 1 + 2y + 2y^2 = 32,$$
$$3y^2 + 2y - 33 = 0,$$
$$(3y + 11)(y - 3) = 0.$$

Since we are looking for non-negative solutions we must have $y = 3$, which means that $z = y^2 = 9$ and $x = \frac{1}{2}(y^2 - 1) = 4$.

(53) *Problem.* We say that there is an *algebraic operation* defined on the closed interval $[0, 1]$ if there is a rule that corresponds to every pair (a, b) of numbers from this interval a new number c from the same interval. We denote it by $c = a \otimes b$.
Find all positive k with the property that there exists an algebraic operation defined on $[0, 1]$ such that for any x, y, z from $[0, 1]$ the following equalities hold:

(i) $x \otimes 1 = 1 \otimes x = x$,
(ii) $x \otimes (y \otimes z) = (x \otimes y) \otimes z$,
(iii) $(zx) \otimes (zy) = z^k(x \otimes y)$.

For all such k define the corresponding algebraic operation(s).
Solution. Let $a \neq 1$ be a non-zero element of $[0, 1]$. For all $r \geq 1$ we see that

$$
\begin{aligned}
a \otimes a^r &= (a \cdot 1 \otimes a \cdot a^{r-1}) & \\
&= a^k(1 \otimes a^{r-1}) & \text{by (iii)} \\
&= a^k(a^{r-1}) & \text{by (i)} \\
&= a^{k+r-1}. & (1)
\end{aligned}
$$

Let s be any real number greater than $\max\{2, k\}$. Then $s - k > 0$. By (ii) we have $(a \otimes a) \otimes a^s = a \otimes (a \otimes a^s)$. Let us examine both sides of this equation separately:

$$
\begin{aligned}
\text{L.H.S.} &= (a \otimes a) \otimes a^s \\
&= a^k \otimes a^s & \text{by (1) with } r = 1 \\
&= (a^k)^k (1 \otimes a^{s-k}) & \text{by (iii)} \\
&= a^{k^2} \cdot a^{s-k} & \text{by (i)} \\
&= a^{k^2 - k + s},
\end{aligned}
$$

$$
\begin{aligned}
\text{R.H.S.} &= a \otimes (a \otimes a^s) \\
&= a \otimes a^{k+s-1} & \text{by (1) with } r = s \geq 2 \\
&= a^{k+(k+s-1)-1} = a^{2k+s-2}
\end{aligned}
$$

$$\text{by (1) with } r = k + s - 1 \geq 1 + k.$$

Since a^x is a one-to-one function, we conclude that the exponents on a in the above expressions must be equal, that is, we have $k^2 - k + s = 2k + s - 2$, which can be rearranged to read $(k-1)(k-2) = 0$. Thus, the only values of k which are possible are $k = 1$ and $k = 2$.

It remains to show that for these values of k we can indeed construct an algebraic operation as defined by (i), (ii), and (iii). Let b and c be any two elements from $[0, 1]$. Clearly, we have exactly one of $b < c$, $b = c$, or $b > c$ true. Thus, we may assume that $c \leq b$ (otherwise, interchange the values of b and c). Then there is a real number $d \in [0, 1]$ such that $bd = c$. Then, if our algebraic operation can be defined, we have

$$b \otimes c = b \otimes bd = b^k (1 \otimes d) = b^k d = b^{k-1} c$$

and a similar computation works for $c \otimes b$. Thus, if $k = 1$, we have $b \otimes c = c \otimes b = c$, the smaller of the two values, that is, $b \otimes c = \min\{b, c\}$. With this definition the properties (i), (ii), and (iii) can all be readily verified.

On the other hand, if $k = 2$, then $b \otimes c = c \otimes b = bc$, that is, the operation $\otimes$ is the ordinary product of two elements. It is again trivial to verify that properties (i), (ii), and (iii) all hold for this definition of $\otimes$. So the only such algebraic operations defined on $[0, 1]$ are the minimum function, and the product.

(54) *Problem.* When a, b, c are the three rational numbers $\frac{1}{2}$, $\frac{1}{3}$ and $\frac{1}{5}$, the value of

$$\frac{1}{(a-b)^2} + \frac{1}{(b-c)^2} + \frac{1}{(c-a)^2} = \frac{1}{(1/6)^2} + \frac{1}{(2/15)^2} + \frac{1}{(3/10)^2}$$

$$= 36 + \frac{225}{4} + \frac{100}{9}$$

$$= \frac{1296 + 2025 + 400}{36}$$

$$= \frac{3721}{36} = \left(\frac{61}{6}\right)^2,$$

is the square of a rational number.

Prove this is no accident, that, in fact, whenever a, b, c are any three different rational numbers, the quantity

$$\frac{1}{(a-b)^2} + \frac{1}{(b-c)^2} + \frac{1}{(c-a)^2}$$

is always the square of a rational number.

Solution. First we will simplify by letting:

$$x = \frac{1}{a-b}, \quad y = \frac{1}{b-c}, \quad z = \frac{1}{c-a}.$$

Then

$$\frac{1}{x} = a-b, \quad \frac{1}{y} = b-c, \quad \frac{1}{z} = c-a,$$

and it is obvious that

$$\frac{1}{x} + \frac{1}{y} + \frac{1}{z} = 0.$$

Clearing fractions yields

$$yz + zx + xy = 0.$$

Now the expression we are dealing with is $x^2 + y^2 + z^2$. Note that

$$(x+y+z)^2 = x^2 + y^2 + z^2 + 2(xy + yz + zx).$$

Since $xy + yz + zx = 0$ we have $x^2 + y^2 + z^2 = (x+y+z)^2$, a perfect square.

(55) *Problem.* The terms a_n of a sequence of positive integers satisfy

$$a_{n+3} = a_{n+2}(a_{n+1} + a_n) \quad \text{for } n = 1, 2, 3, \dots..$$

If $a_6 = 144$, what is a_7?

Solution. The first three terms of the sequence need to be specified in order to start the process. Suppose that

$$a_1 = x, \qquad a_2 = y, \qquad a_3 = z.$$

Then

$$a_4 = z(y + x),$$
$$a_5 = [z(y + x)](z + y),$$
$$a_6 = [z(y + x(z + y))][z(y + x) + z]$$
$$= z^2(y + x)(z + y)(y + x + 1).$$

The presence of the factors $y + x$ and $y + x + 1$ means that two of the factors of $a_6 = 144$ must be *consecutive* positive integers. Since $144 = 2^4 \cdot 3^2$, the positive divisors of 144 (in increasing order) are:

$$1, 2, 3, 4, 6, 8, 9, 12, 16, 18, 24, 36, 48, 72, 144.$$

The only pairs of consecutive integers in this list are $(1, 2)$, $(2, 3)$, $(3, 4)$, and $(8, 9)$. Let us investigate each pair in turn. Since x, y, z are all positive integers, the smallest possible value of the pair $(y + x, y + x + 1)$ is $(2, 3)$. Thus, $(1, 2)$ is ruled out. For the pair $(2, 3)$, we must have $y = x = 1$, whence,

$$144 = z^2(2)(z + 1)(3),$$
$$24 = z^2(z + 1),$$

which has no solution for integer z. For the pair $(3, 4)$ we have

$$144 = z^2(3)(z + y)(4),$$
$$12 = z^2(z + y).$$

Thus, z is either 1 or 2. But $z = 1$ implies that $y = 11$, which is impossible, since $y + x = 3$ and y and x are both positive integers. Therefore, $z = 2$ which forces $y = 1$, and it then follows that $x = 2$ since $y + x = 3$. Thus, we get

$$a_7 = a_6(a_5 + a_4) = 144[2(3)(3) + 2(3)] = 144(24) = 3456.$$

We still, however, need to investigate that last remaining pair $(8, 9)$:

$$144 = z^2(8)(z + y)(9),$$
$$2 = z^2(z + y),$$

which forces $z = 1$ and $y = 1$. Since $y + x = 8$ we also have $x = 7$, whence,

$$a_7 = a_6(a_5 + a_4) = 144[1(8)(2) + 1(8)] = 144(24) = 3456.$$

Thus, in either case, we get $a_7 = 3456$.

(56) *Problem.* Given that a, b, c, d are distinct integers such that $(x - a)(x - b)(x - c)(x - d) = 4$ has a root r which is also an integer, prove that $a + b + c + d = 4r$.

Solution. Without loss of generality, we may assume that $a > b > c > d$, where a, b, c, d are distinct integers. This implies that

$$r - a < r - b < r - c < r - d,$$

where r is an integer satisfying the equation

$$(r - a)(r - b)(r - c)(r - d) = 4.$$

Since all the factors on the left hand side are distinct integers (and increasing in value) and their product is 4, we must have $r - a = -2$, $r - b = -1$, $r - c = 1$ and $r - d = 2$.

Therefore, $(r - a) + (r - b) + (r - c) + (r - d) = 0$, from which it follows that $a + b + c + d = 4r$.

(57) *Problem.* The product of three real numbers is $-\frac{1}{2}$ and the sum of their squares is $\frac{57}{4}$. If the three are taken two at a time and multiplied, the sum of the reciprocals of these products is -7. Determine these three numbers.

Solution. Let the numbers be a, b, and c. Then $abc = -\frac{1}{2}$ and

$$\frac{1}{bc} + \frac{1}{ca} + \frac{1}{ab} = -7.$$

Hence,

$$a + b + c = \frac{7}{2}.$$

Now

$$bc + ca + ab = \frac{1}{2}\left((a + b + c)^2 - (a^2 + b^2 + c^2)\right)$$

$$= \frac{1}{2}\left(\left(\frac{7}{2}\right)^2 - \frac{57}{4}\right)$$

$$= \frac{1}{2} \cdot -\frac{8}{4} = -1.$$

It follows that a, b, and c are roots of the cubic equation

$$x^3 - \frac{7}{2}x^2 - x + \frac{1}{2} = 0.$$

It is easy to check that $x = -\frac{1}{2}$ is a root. Using the Factor Theorem, we get

$$x^3 - \frac{7}{2}x^2 - x + \frac{1}{2} = \left(x + \frac{1}{2}\right)(x^2 - 4x + 1) = 0,$$

which means that the remaining roots are the roots of the quadratic
equation $x^2 - 4x + 1 = 0$, namely $2 \pm \sqrt{3}$. Thus, the three roots
are

$$-\frac{1}{2}, \quad 2 - \sqrt{3}, \quad \text{and} \quad 2 + \sqrt{3}.$$

(58) *Problem.*   Find all distinct pairs (x, y) of real numbers that are
solutions to the equation

$$x^2 - xy + y^2 = x + y.$$

Solution.   We first rewrite the equation as

$$y^2 - y(x + 1) + x^2 - x = 0.$$

Using the quadratic formula, we get

$$y = \frac{x + 1 \pm \sqrt{(x + 1)^2 - 4(x^2 - x)}}{2} = \frac{x + 1 \pm \sqrt{1 + 6x - 3x^2}}{2}.$$

In order to obtain real solutions we need the discriminant of the
quadratic, namely $1 + 6x - 3x^2$ to be non-negative. That is,

$$0 \le 1 + 6x - 3x^2 = 1 - 3(x^2 - 2x) = 4 - 3(x^2 - 2x + 1) = 4 - 3(x - 1)^2,$$

which means that $(x - 1)^2 \le \frac{4}{3}$. Since we need x to be an integer,
we conclude that $(x - 1)^2 = 0$ or 1, which means x can only be one
of 0, 1, or 2. Solving this equation with $x = 0$ yields $y = 0$ or 1;
with $x = 1$ yields $y = 0$ or 2; and with $x = 2$ yields $y = 1$ or 2.
Thus, the solution set is $(x, y) = (0, 0)$, $(1, 0)$, $(0, 1)$, $(1, 2)$, $(2, 1)$,
and $(2, 2)$.

(59) *Problem.*   Find all possible real solutions of the following system
of equations:

$$z^x = y^{2x},$$
$$2^z = 2(4)^x,$$
$$x + y + z = 16.$$

Solution.   If $x = 0$ then the first equation is solved (as long as
we insist that y and z are non-zero), and this yields the solution
$(x, y, z) = (0, 15, 1)$ for the system.

Now assume that $x \ne 0$. The first equation then gives $z = \pm y^2$ and
the second $z = 2x + 1$. Substituting first $x = \frac{1}{2}(z - 1)$ and then
$z = y^2$ into the third equation generates the quadratic equation

$$3y^2 + 2y - 33 = (3y + 11)(y - 3) = 0.$$

Thus, $y = 3$ or $y = -11/3$, and we have the possible solutions $(4, 3, 9)$ and $\left(\frac{56}{9}, -\frac{11}{3}, \frac{121}{9}\right)$.

Substituting first $x = \frac{1}{2}(z - 1)$ and then $z = -y^2$ into the third equation generates the quadratic equation

$$3y^2 - 2y + 33 = 0,$$

which has no real roots.

Therefore, the complete solution set for (x, y, z) is

$$(0, 15, 1), \quad (4, 3, 9), \quad \text{and} \quad \left(\frac{56}{9}, -\frac{11}{3}, \frac{121}{9}\right).$$

(60) *Problem.* Find a necessary and sufficient condition on the coefficients a, b, and c in the quadratic equation

$$ax^2 + bx + c = 0,$$

where $a \neq 0$, in order that one of its roots is a square of the other root.

Solution. Suppose first that one of the roots is a square of the other root, say the roots are r and r^2. Then $r^2 + r = -b/a$ and $r^3 = r^2 \cdot r = c/a$. Now

$$(r^2 + r)^3 = r^3(r+1)^3 = r^3(r^3 + 3r^2 + 3r + 1) = r^3\left(r^3 + 3(r^2 + r) + 1\right),$$

or

$$\left(-\frac{b}{a}\right)^3 = \frac{c}{a}\left(\frac{c}{a} - \frac{3b}{a} + 1\right), \tag{1}$$

which is equivalent to

$$b^3 + ca(c + a) = 3abc. \tag{2}$$

On the other hand, suppose that equation (2) holds, and let the roots of the $ax^2 + bx + c = 0$ be r and s. Substituting $-b/a = r + s$ and $c/a = rs$ in the equivalent condition (2) gives

$$(r + s)^3 = rs\left(rs - 3(r + s) + 1\right),$$

which can rearranged as

$$(r^2 - s)(s^2 - r) = 0.$$

Thus, one of the roots is the square of the other.

Therefore, a necessary and sufficient condition is that

$$b^3 + ca(c + a) = 3abc.$$

(61) *Problem.*   Given that a, b, x, and y are real numbers such that:

$$a + b = 23,$$
$$ax + by = 79,$$
$$ax^2 + by^2 = 217,$$
$$ax^3 + by^3 = 691,$$

determine $ax^4 + by^4$.

Solution. Let us number the equations above from 1 to 4. Multiplying equation 2 by $x + y$ yields:

$$(ax + by)(x + y) = ax^2 + bxy + aa^+by^2$$
$$= (ax^2 + by^2) + xy(a + b)$$

or $79(x + y) = 217 + 23xy.$ (1)

Similarly multiplying equation 3 by $x + y$ yields:

$$(ax^2 + by^2)(x + y) = ax^3 + ax^2y + bxy^2 + by^3$$
$$= (ax^3 + by^3) + xy(ax + by)$$

or $217(x + y) = 691 + 79xy.$ (2)

Setting $w = x + y$ and $z = xy$, equations 1 and 2 can be written as the linear system:

$$217w - 79z = 691$$
$$79w - 23z = 217,$$

which has solution $w = 1$ and $z = -6$. Thus, we have $x + y = 1$ and $xy = 6$. This yields exactly two solutions: $(x, y) = (3, -2)$ or $(-2, 3)$. Since the roles played by x and y are interchangeable, we may choose one of the above solutions without any loss of generality, say $(x, y) = (3, -2)$. With these values of x and y we compute $a = 25$ and $b = -2$, from which we obtain $ax^4 + by^4 = 1993$.

(62) *Problem.*   Show that

$$\left(\sqrt{5} + 2\right)^{\frac{1}{3}} - \left(\sqrt{5} - 2\right)^{\frac{1}{3}}$$

is a rational number and find its value.

Solution.   Let $k = a - b$ where $a = (\sqrt{5} + 2)^{\frac{1}{3}}$ and $b = (\sqrt{5} - 2)^{\frac{1}{3}}$. Then

$$k^3 = a^3 - 3a^2b + 3ab^2 - b^3 = a^3 - b^3 - 3ab(a - b)$$
$$= (\sqrt{5} + 2) - (\sqrt{5} - 2) - 3\left((\sqrt{5} + 2)(\sqrt{5} - 2)\right)^{\frac{1}{3}} k$$
$$= 4 - 3 \times 1^{\frac{1}{3}} \times k = 4 - 3k.$$

This cubic can be factored as $(k-1)(k^2+k+4) = 0$, whence $k = 1$, since the quadratic part cannot be factored in the reals.

(63) *Problem.* Find all solutions (x, y) to the system of equations:

$$x + y + \frac{x}{y} = 19,$$

$$\frac{x(x + y)}{y} = 60.$$

Solution. First we should notice that both equations contain the terms $x + y$ and x/y, and matters might be simplified if we use two new variables, say z and w, to represent these expressions. This yields the system:

$$w + z = 19,$$

$$wz = 60.$$

Solving the first for w and substituting into the second yields the quadratic:

$$z^2 - 19z + 60 = 0,$$

which has solutions $z = 4$ or 15, with the corresponding values for w being 15 and 4. This then yields two new systems for x and y: the system

$$\left.\begin{array}{l} x + y = 4, \\ \dfrac{x}{y} = 15, \end{array}\right\} \quad \text{or the system} \quad \left\{\begin{array}{l} x + y = 15, \\ \dfrac{x}{y} = 4. \end{array}\right.$$

In the first system, solving the second equation for x and substituting into the first yields $y = 1/4$ (and $x = 15/4$). In the second system the same approach yields $y = 3$ (and $x = 12$).

Thus, the complete solution for (x, y) is:

$$\left(\frac{15}{4}, \frac{1}{4}\right) \quad \text{and} \quad (12, 3).$$

(64) *Problem.* Solve

$$\sqrt[3]{13x + 37} - \sqrt[3]{13x - 37} = \sqrt[3]{2}.$$

Solution. Cubing both sides of the given equation yields

$$(13x + 37) - 3\sqrt[3]{(13x + 37)^2(13x - 37)}$$

$$+ 3\sqrt[3]{(13x + 37)(13x - 37)^2} - (13x - 37) = 2,$$

$$74 - 3\sqrt[3]{(13x + 37)(13x - 37)}(\sqrt[3]{13x + 37} - \sqrt[3]{13x - 37}) = 2,$$

$$72 - 3\sqrt[3]{169x^2 - 1369}(\sqrt[3]{2}) = 0,$$

$$\sqrt[3]{2}\sqrt[3]{169x^2 - 1369} = 24.$$

Cubing this last equation, we get

$$169x^2 - 1369 = 6912,$$

which has solution $x = \pm 7$, both of which solve the original equation.

(65) *Problem.* Determine the number of real solutions a to the equation

$$\left[\frac{1}{2}a\right] + \left[\frac{1}{3}a\right] + \left[\frac{1}{5}a\right] = a.$$

Here, if x is a real number, then $[x]$ denotes the greatest integer that is less than or equal to x.

Solution. Since the left hand side of the given equation is a sum of integers, the right hand side, namely a, must also be an integer. Let $a = 30q + r$ where r is an integer satisfying $0 \leq r \leq 29$. Then

$$\left[\frac{1}{2}a\right] = \left[\frac{1}{2}(30q + r)\right] = 15q + \left[\frac{r}{2}\right],$$

$$\left[\frac{1}{3}a\right] = \left[\frac{1}{3}(30q + r)\right] = 10q + \left[\frac{r}{3}\right],$$

$$\left[\frac{1}{5}a\right] = \left[\frac{1}{5}(30q + r)\right] = 6q + \left[\frac{r}{5}\right].$$

The given equation then translates to

$$15q + \left[\frac{r}{2}\right] + 10q + \left[\frac{r}{3}\right] + 6q + \left[\frac{r}{5}\right] = 30q + r,$$

$$q = r - \left[\frac{r}{2}\right] - \left[\frac{r}{3}\right] - \left[\frac{r}{5}\right].$$

For each value of r, $0 \leq r \leq 29$, we get exactly one value for q, and we also get a different value of $a = 30q + r$ for each value of r. Since there are 30 different values of r, we have 30 solutions for a. For those who are curious, we can compute q and a for each value of r from 0 to 29 and determine all 30 solutions for a. They are (in order of increasing values of r): 0, 31, 32, 33, 34, 35, 6, 37, 38, 39, 10, 41, 12, 43, 44, 15, 16, 47, 18, 49, 20, 21, 22, 53, 24, 25, 26, 27, 28, and 59.

(66) *Problem.* Solve in integers for x and y:

$$6(x! + 3) = y^2 + 5.$$

Solution. The equation can be simplified to

$$6x! + 13 = y^2.$$

For $x \geq 5$, we note that $x!$ is a multiple of 10; in this case the left hand side of the equation ends in 3 (in decimal notation). However, the units digit of a perfect square is one of 1, 4, 5, 6, 9. Thus, $x \leq 4$. The factorial notation only makes sense for x non-negative. Therefore, it remains for us to test the 5 possible values of x, namely 0, 1, 2, 3, and 4. This yields the solutions $(x, y) = (2, \pm 5)$ and $(x, y) = (3, \pm 7)$.

(67) *Problem.* Solve the following equation:

$$|x + 1| - |x| + 3|x - 1| - 2|x - 2| = x + 2.$$

Solution. At least one of the four absolute value expressions is unravelled differently on either side of the numbers -1, 0, 1, and 2. Thus, we shall first look for solutions lying in the interval from 2 to ∞, then in the interval from 1 to 2, then 0 to 1, then -1 to 0, and finally from $-\infty$ to -1.

Let $x \geq 2$. Then $x + 1 > 0$, $x > 0$, $x - 1 > 0$, and $x - 2 \geq 0$, which implies that

$$|x + 1| = x + 1,$$
$$|x| = x,$$
$$|x - 1| = x - 1,$$
$$|x - 1| = x - 2.$$

Thus, the given equation becomes:

$$x + 1 - x + 3(x - 1) - 2(x - 2) = x + 2,$$

which is an identity. Thus, all real numbers greater than or equal to 2 are solutions of the given equation.

Now let $1 \leq x < 2$. Then $x + 1 > 0$, $x > 0$, $x - 1 \geq 0$, and $x - 2 < 0$, which implies that

$$|x + 1| = x + 1,$$
$$|x| = x,$$
$$|x - 1| = x - 1,$$
$$|x - 1| = -(x - 2).$$

In this case the given equation becomes:

$$x + 1 - x + 3(x - 1) + 2(x - 2) = x + 2.$$

This simplifies to $4x = 8$, or $x = 2$, which lies outside our current interval. Consequently there is no additional solution of the given equation found between 1 and 2.

Let $0 \leq x < 1$. Then, arguing as above, we see that $|x+1| = x+1$, $|x| = x$, $|x - 1| = -(x - 1)$, and $|x - 2| = -(x - 2)$. The given equation then becomes:

$$x + 1 - x - 3(x - 1) + 2(x - 2) = x + 2.$$

This simplifies to $-x = x+2$, or $x = -1$. But this value lies outside the interval from 0 to 1 which was used to set up the equation; it must therefore be discarded. Thus, there is no additional solution to the given equation found between 0 and 1.

Let $-1 \leq x < 0$. Then, arguing as above, we obtain that $|x+1| = (x+1)$, $|x| = -x$, $|x-1| = -(x-1)$, and $|x-2| = -(x-2)$. The given equation then becomes:

$$x + 1 + x - 3(x - 1) + 2(x - 2) = x + 2.$$

This simplifies to $x = x+2$, which is impossible; thus, there are no solutions found between -1 and 0.

Finally let $x < -1$. Then, as above, $|x + 1| = -(x + 1)$, $|x| = -x$, $|x - 1| = -(x - 1)$, and $|x - 2| = -(x - 2)$. The given equation then becomes:

$$-(x + 1) + x - 3(x - 1) + 2(x - 2) = x + 2.$$

This simplifies to $-x - 2 = x + 2$, or $x = -2$, which value does indeed lie in the interval we are considering. Therefore, the given equation is satisfied by -2 and all real numbers greater than or equal to 2.

Chapter 5

Logic, Games, Puzzles and Amusements in Math

(68) *Problem.*   You have just found yourself on the magical island of Consistency, where there live 3 tribes (all of whom use English for communication), called Whites, Reds, and Blues.

The 3 tribes are characterized in the following way:

– Whites always tell the truth,

– Reds always lie,

– Blues alternate between a lie and the truth, but one never knows with which they start.

These tribes have no distinguishing physical characteristics.

After walking a ways you encounter 3 persons you know (somehow) are each from a different tribe. The names of the 3 persons are Mr. White, Mr. Red, and Mr. Blue. You then have the following conversation with Mr. Blue:

(a) "To which tribe does Mr. White belong?"
(b) "White, of course."
(c) "And Mr. Red?"
(d) "Red!"
(e) "And then you must be Blue?"
(f) "Right you are!"

Is Mr. Red really a Red? If not, what is he?

Solution. If Mr. Blue is really a Blue, then his statements must be respectively either TFT or FTF. However, if two of them are true, the third must also be true (since there is one of each of the 3 tribes present). Thus, TFT is eliminated.

Therefore, if Mr. Blue is a Blue, his statements must be FTF. But he says in his third answer that he is a Blue, which would be true if he were indeed a Blue. Therefore, Mr. Blue is <u>not</u> a Blue.

If he is a White, he always tells the truth, but he says in his third answer that he is a Blue, which is another contradiction.

Therefore, Mr. Blue is a Red. Since he is a Red, he always lies. Thus, Mr. White is not a White. Since Mr. Blue is a Red, Mr. White must be a Blue.

This leaves Mr. Red to be a White.

(69) *Problem.*   On a certain day, three sportscasters predicted the winners in the Bush League as follows:

> I: New York, Chicago, Detroit, Pittsburgh;
> II: Cleveland, Boston, Detroit, Chicago;
> III: Detroit, Pittsburgh, Philadelphia, Boston.

The league consisted of the teams: New York, Boston, Kansas City, Philadelphia, Cleveland, Pittsburgh, Detroit, and Chicago. Who played whom that day?

Solution. If a sportscaster picks four teams as the likely winners of four different games on a given day, then it is safe to assume that those four teams are participating in different games. Thus, we may conclude that Detroit does not play any other team mentioned in the predictions. The only team not mentioned was Kansas City. Therefore, Detroit must have played Kansas City. Let us disregard these two teams and concentrate on the remaining six teams. We may then conclude that Boston played New York, because Boston is predicted as the winner by either II or III together with each of the other four teams. Similarly, we see that Chicago played Philadelphia by considering I and II, which leaves Pittsburgh playing Cleveland.

That is,

> Detroit   vs Kansas City,
> Boston   vs New York,
> Chicago   vs Philadelphia,
> Cleveland vs Pittsburgh.

(70) *Problem.*   Four men, Adams, Bates, Clark, and Douglas, were speaking of their wives. They were not well acquainted and the statements they made (listed below) were not all accurate. In fact, the only thing which is certain is that each statement in which a man mentions his wife's name is correct.

(a) Adams: Dorothy is Jean's mother.
(b)   I have never met Patricia.
(c) Bates: Clark's wife is either Dorothy or Patricia.
(d)   Jean is the oldest.
(e) Clark: Patricia is Adams' wife.
(f)   Dorothy is Jean's older sister.
(g) Douglas: Margaret is my daughter.
(h)   Dorothy is older than my wife

What is the given name of each man's wife?

Solution. From Douglas's statements, we may conclude that he is not married to either Margaret or Dorothy. By similar reasoning, we may also conclude that Adams is not married to Patricia, and neither is Clark. Now suppose that Clark is married to Margaret. Then the first statement made by Bates is false, from which we conclude that Bates is not married to either Dorothy or Patricia. Since Bates cannot be married to Margaret either (she is married to Clark by assumption), we conclude that Bates is married to Jean, which means that Jean is the oldest. Since Douglas could only be married to Patricia or Jean, it follows that he is married to Patricia, leaving Dorothy as the wife of Adams. However, this contradicts the fact that Jean is the oldest. Therefore, Clark is NOT married to Margaret.

This means that Clark is married to either Dorothy or Jean, and we conclude that Clark's second statement is true; that is, Dorothy is Jean's older sister. This further tells us that Bates is NOT married to Jean, and that Adams is NOT married to either Dorothy or Jean. Thus, Adams is married to Margaret. This means that Bates is married to either Dorothy or Patricia, which means that his first statement is true, further implying that Clark is married to Dorothy or Patricia. Combining this with the first sentence in this paragraph tells us that Clark is married to Dorothy. Our solution is then:

Margaret Adams, Patricia Bates, Dorothy Clark, and Jean Douglas.

(71) *Problem.* Four men, one of whom was known to have committed a certain crime, made the following statements when questioned by the police:

Archie:	Dave did it.	(1)
Dave:	Tony did it.	(2)
Gus:	I didn't do it.	(3)
Tony:	Dave lied when he said I did it.	(4)

(a) If only one of these four statements is true, who was the guilty man?

(b) If only one of them is false, who was the guilty man?

Solution. Let us refer to the statements by number. First notice that statements #2 and #4 cannot both be true, nor can they both be false.

(a) If only one of the statements is true, then both statements #1 and #3 are false. Thus, statement #3 implies that Gus is guilty.

(b) If only one of the statements is false, then both statements #1 and #3 are true. Therefore, statement #1 tells us that Dave is guilty.

(72) *Problem.* "Do you play billiards? Care to have a game?" asked Huntingdon of the new member at the Town Club.

"Yes, I play," replied McClintock, "But I'm rather a duffer. My friend Chadwick gives me 25 points in 100, and then we play about even."

"Well, I'm perfectly willing to give you a proper handicap. I give Chadwick 20 points in 100. Now let's see how many points should I give you?"

What is the correct answer, assuming that the stated handicaps are fair?

Solution. If Huntingdon gives Chadwick 20 points in 100, then Chadwick is expected to make 80 while Huntingdon makes 100, thus, scoring points at $\frac{4}{5}$ of Huntingdon's rate. Similarly, McClintock scores at $\frac{3}{4}$ the rate of Chadwick. Then, McClintock should score at $\frac{4}{5} \times \frac{3}{4} = \frac{3}{5}$ the rate of Huntingdon. Thus, Huntingdon should give McClintock 40 points in 100.

(73) *Problem.* Three counters A, B, C are coloured red, white, and blue, but not necessarily respectively. Of the following statements, only one is true:

$$A \text{ is red}; \quad B \text{ is not red}; \quad C \text{ is not blue}.$$

What colour is each counter? Justify your answer.

Solution. If A is red, then B is not red, which makes two statements correct, which is impossible. Thus, we conclude that A is not red. Then either B or C is red. If C is red, then again B is not red and two statements are true, which is impossible.

Therefore, C is not red, implying that B is red. Thus, the only true statement must be the last one. This means that C is white, and it follows that A is blue. Conclusion:

$$A \quad \text{blue}$$
$$B \quad \text{red}$$
$$C \quad \text{white}$$

(74) *Problem.* A says "B is a liar, or C is a liar, or both are liars". B says "A is a liar".
C says "A is a liar and B is a liar".

Who is telling the truth? Justify your answer.
Solution. If C is telling the truth, then B is a liar. But B and C agree on A. Thus, C is a liar. Consequently, A's statement is true. Therefore, A is telling the truth, and B and C are lying.

(75) *Problem.* A set of balls is given. Each ball is coloured red or blue, and there is at least one of each colour. Each ball weighs either 1 pound or 2 pounds, and there is at least one of each weight.
Prove that there are two balls having different weights and colours.
Solution. There is at least one red ball and one blue ball. Consider one of each. If the two balls have different weights, we are done. If they both have the same weight, choose a third ball with a different weight. The colour of the third ball must differ from one of the first two balls, and the weights are definitely different.
Thus, the statement is proved.

(76) *Problem.* Four members of my club – Messrs. Albert, Charles, Frederick, and Richard – have recently been knighted. Now their friends have to learn their given names.
These are a little worrying: for it transpires that the surname of each of the four knights is the given name of one of the others.
Richard is not the given name of the member whose surname is Albert. There are three of the knights related as follows: the given name of the member whose surname is Frederick is the surname of the member whose given name is the surname of the member whose given name is Charles.

What is Mr. Richard's given name?

Solution. Consider the last clue first. This implies that the three persons are:

X	Frederick
Y	X
Charles	Y

where column 1 contains the given names and column 2 contains the surnames. It is clear from the clues that X and Y must be different from Charles and Frederick.

Thus, the fourth person must be Frederick Charles. Since Richard Albert is <u>not</u> permitted, we must have that X = Richard and Y = Albert.

Therefore, the four persons are:

Richard	Frederick
Albert	Richard
Charles	Albert
Frederick	Charles

(77) *Problem.* Three men named Lewis, Miller, and Nelson fill the positions of accountant, cashier, and clerk in the leading department store in Spuzzum.

  (a) If Nelson is the cashier, Miller is the clerk.
  (b) If Nelson is the clerk, Miller is the accountant.
  (c) If Miller is not the cashier, Lewis is the clerk.
  (d) If Lewis is the accountant, Nelson is the clerk.

What is each man's job?

Solution. Let us first assume that the four statements above are numbered from 1 to 4. Then notice first that statements 4 and 2 taken together imply that Lewis is not the accountant. On the other hand if Lewis is the cashier, statement 3 says that Lewis must be the clerk. Consequently, Lewis cannot be the cashier either, but must be instead the clerk. This fact taken together with statement 1 shows that Nelson cannot be the cashier. Thus Nelson must be the accountant, which leaves Miller as the cashier.

(78) *Problem.* In Italy, in the 16th century, there were only two families who made coffins: Bellini and his sons on one hand, and Fellini and his sons on the other hand.

Each time that Bellini or one of his sons finished a coffin, they engraved a true inscription on it. Fellini and his sons engraved on their coffins a false inscription.

Recently a coffin was discovered bearing the inscription: "This coffin has not been made by a son of Bellini".

Could the coffin have been made in 16th century Italy? If so, who made it? Explain.

Solution. Since Fellini and his sons always engrave a false inscription on their coffins, they could not possibly have made this particular coffin. Also, since Bellini and his sons always engrave a true inscription on their coffins, no son of Bellini could have made this coffin.

However, there is no way to rule out the possibility that Bellini himself made the coffin. Thus, it could have been made in 16th century Italy, but only if Bellini himself made it.

(79) *Problem.* A team of workers was assigned to cut the hay in two meadows, one of which is twice the size of the other. The team spent half a day on the larger meadow, then split in half. One group continued to work in the larger meadow, and finished it by evening. The second group spent the rest of the day working in the smaller meadow, but by evening there still remained a small section left uncut. A single worker was able to finish that part in another full day's work.

How many workers were in the original team?

Solution. Let n be the number of workers in the original team and let us assume that all workers work equally well. The number of 'workdays' to clear the larger field can be seen to be $\frac{1}{2}n + \frac{1}{4}n$, which is $\frac{3}{4}n$ workdays, and the number of workdays to clear the smaller field is $\frac{1}{4}n + 1$. Since the larger field is twice the size of the smaller field, we conclude

$$\frac{3}{4}n = 2\left(\frac{1}{4}n + 1\right),$$

whence $n = 8$.

Thus, there were 8 men in the original team.

(80) *Problem.* Fred and Frank are two fitness fanatics on a run from A to B. Fred runs half the way and walks the other half. Frank runs for half the time and walks for the other half. They both run and walk at the same speeds. Who finishes first? (You may assume that they run faster than they walk!!)

Solution. Although the problem states that you may assume that both men run faster than they walk, you will get the same conclusion even if you assume the exact opposite! The only assumption that is necessary is that they run and walk at different speeds. Let w be the speed at which they both walk and let r be the speed at which they both run. Further let D be the distance from A to B. We may assume that the total amount of time that each walks is continuous, and the same for the total amount of time for running. Furthermore, let us suppose that they both start running. If we assume that running is faster than walking, then Frank, who runs and walks half the time, will spend more time running.

Thus, he will still be running for a period of time after Fred has started to walk. Since they walk at the same speed as each other, it is only at this time that Frank will forge into the lead. When he stops running he will be in the lead and will maintain that lead since they walk at the same speed.

Therefore, Frank will finish first.

(81) *Problem.* Luke and Slim have only one horse between them. Luke rides a certain time and then ties up the horse for Slim, who has been walking. Meanwhile, Luke walks on ahead. They proceed in this way, alternately walking and riding. If they walk 4 miles per hour and ride 12 miles per hour, what part of the time is the horse resting?

Solution. We must first observe that horse and rider travel 3 times as fast as a man walking. Thus, in any unit of time the horse with rider will travel 3 times as far as a man walking. Since they each walk the same speed, for any distance covered they will each walk half and ride half. Think of the amount of time taken to ride half the distance as a 'unit' of time. During the unit of time that Slim is riding half the distance, Luke is walking (but only one sixth of the total distance), and vice versa. Thus, during these two units of time, they have each covered $\frac{2}{3}$ of the total distance. Since the remaining $\frac{1}{3}$ of the total must still be walked, and it would require 2 units of time, we see that it takes a total of 4 units of time to cover the total distance, of which only two units are involved in riding. Therefore, the horse resting one half of the time.

(82) *Problem.* In the game of Sixty, two players take turns adding an integer from 1 to 7 to an accumulated sum; the person who reaches a sum of 60 wins. For example, the first player could start

with 3, the second might then add 7 to make a sum of 10; now the first player might add 6 to make 16, and the second might add 5 to make 21, and so on until 60 is reached. Who ought to win the game: the first player or the second player? What strategy will guarantee a win? How would you answer these questions if the goal was to reach 64 instead of 60?

Solution. Let us begin by recognizing that when the sum is 53 or greater, the player who is about to play can force a win. This means that he has been given a winning position. However, if the total is 52, it is easy to see that the player who is about to play cannot win. Thus, it should be the strategy to bring the total to 52 when it becomes the opponent's turn to play. This means that we can logically replace the winning sum of 60 by the sum of 52, which guarantees a win for the person reaching it. By a similar analysis we can see that 44, 36, 28, 20, 12, and finally 4 are also winning positions. Consequently, the first person should pick 4 and for every choice of number made by the second person, simply choose the difference between 8 and the first player's choice. This will force the sum to fall into the list of winning positions enumerated above, which will give the first player a win every time.

On the other hand, if the winning position is to be 64 instead of 60, then the set of decreasing winning positions are all the multiples of 8 from 64 down to 0 (use an argument similar to above). This means that when the first person plays he is forced to give up a winning position! Thus, if the goal is 64 the second player is the winner.

(83) *Problem.* Below is a multiplication problem in which all the digits but one have been removed. Can you reconstruct the problem, given that the 1 shown is the only occurrence of the digit 1 in the multiplication?

$$
\begin{array}{r}
- \, - \\
\times \; - \, - \\
\hline
- \, - \\
- \, 1 \\
\hline
- \, - \, -
\end{array}
$$

Solution. Let us place letters into each position to ease discussion. Our problem looks like:

$$A B$$
$$\times C D$$
$$\overline{ E F}$$
$$\underline{G 1}$$
$$?\ ?\ ?$$

where each letter represents a digit (not necessarily distinct) and the three question marks represent three (not necessarily distinct) digits. Since $B \times C$ ends in 1 and neither can be 1, we conclude that $(B, C) = (3, 7)$, $(7, 3)$, or $(9, 9)$. If C is 9, then $AB \times C$ must be three digits in length, but we see that it is $G1$. This contradiction rules out $(9, 9)$ as a possibility. Consider $(3, 7)$ as a possible value of (B, C); then $A3 \times 7$ has only two digits, which forces A to be 1, which is impossible. Therefore, $B = 7$ and $C = 3$. We still must have $A > 1$. But $A > 3$ implies that $G1$ has more than two digits. Therefore, $A = 2$, which in turn implies that $G = 8$.

Now $D > 1$. If $D > 4$, then EF would be larger than two digits, and if $D = 3$, then F would be 1, which is also impossible. Thus, we conclude that $D = 2$, which yields

$$27$$
$$\times 32$$
$$\overline{54}$$
$$\underline{81}$$
$$864$$

(84) *Problem.* A certain physicist, who is always in a hurry, walks up an upgoing escalator at the rate of one step per second. Twenty steps bring him to the top. Next day he goes up at two steps per second, reaching the top in 32 steps. How many steps are there in the escalator?

Solution. Let s be the number of steps on the escalator and let r be the rate of the escalator in steps per second. In the time taken for the physicist to walk 20 steps on the escalator (that is, 20 seconds), $s - 20$ steps have gone "over the top". But the number of steps which have gone "over the top" in t seconds is rt. Therefore,

$$s - 20 = 20r.$$

Similarly,

$$s - 32 = 16r,$$

by considering the performance on the second day. Solving these two equations simultaneously, we obtain $s = 80$ steps.

(85) *Problem.*   The figure below shows five rings, four of brass and one of rubber, lying almost flat on a table. Naturally the rubber ring is quite flexible, while the brass rings are flat and rigid. All we ask you to do is to say which of the rings is made of rubber.

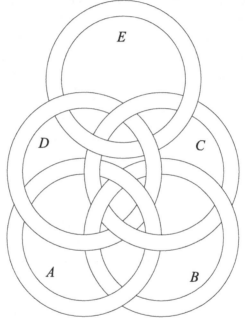

Solution.   The solution depends on the fact that if there are three rings which interlock in the manner shown by the diagram below, then one of the rings must be flexible if they are all to lie flat. Note that, in moving in a clockwise direction, each ring goes over each successive ring.

In this puzzle, C is over D which is over E which is over C, so one of these rings must be flexible. Similarly, one of D, B, and C is rubber, and so is one of A, B, and C. The only common ring is C, which is therefore the flexible ring.

(86) *Problem.*   Which of the following poker hands is the best? (That is, on which hand would I be willing to bet the most money?) Which hand is the worst? Which hands are of equal strength? The game is being played with an ordinary 52-card pack. There are no

wild cards. (AS means ace of spades, etc.)

(i) AS AH AD KS KH
(ii) AS AH AD QS QC
(iii) AS AH AD QS QH
(iv) AS AH AD 6S 6C
(v) AS AH AD 3S 3C

Solution. Since the game is being played with only one 52-card deck of cards with no wild cards, it is quite clear that no two of the above hands could possibly occur on the same deal. Thus, the only way to compare their relative values is to determine the number of hands which could beat them. Since they are all "full houses with aces", there is no other full house which will beat any of them. Also, each of them will lose to any hand with 4-of-a-kind and there are precisely 11 such hands possible. Therefore, the difference, if any, must be in the number of straight flushes which they lose to; since they will lose to any straight flush, we need only count the number of straight flushes possible given that we are looking at one of the above hands.

In hand (i) there are still 31 possible winning straight flushes; in hand (ii) there are only 28 possible winning straight flushes; in hand (iii) there are 29 winning possibilities; in hand (iv) there are only 23 winning possibilities; and in hand (v) there are 29 still available.

Therefore, the best hand is hand (iv).

(87) *Problem.* Suppose n persons attend a meeting. Each attendee is acquainted with exactly eight other attendees. Furthermore, suppose that each pair of attendees who are acquainted with each other has four acquaintances in common among the attendees at the meeting, and each pair of attendees who are NOT acquainted with each other has only two acquaintances in common. What are possible values of n?

Solution. Such a situation cannot occur. Suppose instead that such a meeting is possible. Let x be an attendee of the meeting, and let x be acquainted with $x_1, x_2, \ldots, x_8$. Since x is acquainted with x_8, they have four acquaintances in common among $x_1, x_2, \ldots, x_7$, say x_1, x_2, x_3, x_4. In particular, x_7 and x_8 are NOT acquainted. Since x is acquainted with x_7, they also have four acquaintances in common among $x_1, x_2, \ldots, x_6$.

Thus, at least two must be among x_1, x_2, x_3, x_4, say x_1 and x_2. But now x_7 and x_8, who are NOT acquainted, have x, x_1, and x_2 as common acquaintances, which is too many, a contradiction. Therefore, no such meeting can occur.

(88) *Problem.* Mr. Smith was expecting a package, so he tacked the following note to his door:

"The doorbell on my apartment doesn't work. If you want leave a package, use the knocker. If nobody answers, ring the doorbell of my neighbour next door. If no one answers there, try my sister's at apartment 2."

When the postman arrived with the package, he found the note had fallen to the floor, and he could not tell which door it had been pinned to. There were 6 apartment doors side by side, numbered 1 to 6. Apartments 1, 3, 4, and 5 were equipped with knockers and all apartments except 4 had doorbells.

Seeing this the postman was able to deliver the package correctly. Which apartment is Mr. Smith's?

Solution. Since Mr. Smith claims to have a doorbell and a knocker, it is clear that only apartments 1, 3, and 5 qualify.

One may also deduce from the note that his neighbour and his sister are not the same person. We may thus eliminate apartment 1.

If apartment 3 was Mr. Smith's, his neighbour would have to be in apartment 4, which has no doorbell, contrary to the statement in the note.

The only alternative is that Mr. Smith lives in apartment 5, and it can be checked that all the conditions are satisfied in this case.

(89) *Problem.* "How did you both do in the exam?" asked Walt, when the children came home.

"Not so good," replied Ann. "I got a third of them wrong."

"It was tough," Bill told his father. "I got five of them wrong, but between us two we answered three-quarters of the questions right." How many did Ann answer correctly?

Solution. Each question on the test falls into one of 4 categories:

1. Both Ann and Bill answered the question correctly.
2. Ann answered it correctly and Bill did not.
3. Bill answered it correctly and Ann did not.
4. Both Ann and Bill answered it incorrectly.

Let x, y, z, w be the number of questions in each of these four categories, respectively.

Then the total number of questions on the exam is $x + y + z + w$.
From Ann's statement we conclude that

$$x + y = 2(z + w), \tag{1}$$

and from Bill's statement we see that

$$y + w = 5 \tag{2}$$

$$\text{and} \quad x + y + z = 3w. \tag{3}$$

From (2) we have $w = 5 - y$. When we substitute this into (1) and
(3) we obtain

$$x + 3y - 2z = 10$$

$$\text{and} \quad x + 4y + z = 15,$$

whence, we obtain

$$3x + 11y = 40.$$

Since both x and y are positive integers, we must have $x = 6$ and
$y = 2$. From this we see that $w = 3$ and $z = 1$. Thus, there were
12 questions and Ann answered 8 correctly.

(90) *Problem.*   Eight numbered cards lie face down on a table in the
relative positions shown in the diagram below.

Of the eight cards:

1. There is at least one Queen.
2. Every Queen lies between two Kings.
3. There is at least one King between two Jacks.
4. No Jack borders on a Queen.
5. There is exactly one Ace.
6. No King borders on the Ace.
7. At least one King borders on a King.

8. Each card is either a King, a Queen, a Jack, or an Ace.

Which one of the eight numbered cards is the Ace?

Solution. From statements 1 and 2 we conclude that one of the cards numbered 3, 4, or 6 is a Queen surrounded by two Kings. Similarly, statement 3 implies that one of these same cards must be a King surrounded by two Jacks.

If card 3 is a Queen, then by the above arguments and statement 4, we conclude that cards 1 and 6 must be Jacks, whence, statement 7 must be false. Thus, card 3 is not a Queen.

If card 6 is a Queen, then (by statement 4) the only cards which could be Jacks are cards 1, 2, and 3, which would make statement 3 false. Thus, card 6 is not a Queen.

Therefore, card 4 is a Queen, cards 1 and 6 are Kings and cards 5 and 7 are Jacks. Statement 6 prohibits card 8 being the Ace. So, either card 2 or 3 is the Ace. This implies that neither of these cards can be a King (statement 6). Neither card 2 nor 3 can be Queen by statement 2, and by statement 4, card 3 cannot be a Jack. Thus, the only possibility for card 3 is the Ace (statement 8).

Further, to complete the solution in a consistent manner requires that card 2 be a Jack and card 8 be a King.

(91) *Problem.* Given the small chess board (3×4) shown below with a White Queen and Black King as shown, find a way in four moves or less for White to force the Black King into the upper right square (3d).

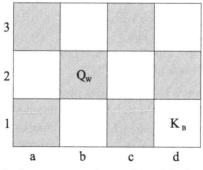

Solution. Each person needs to play with the problem to become convinced that there is no other way to proceed. It is interesting that the solution never involves putting the king in check!

	QUEEN			KING	
1.	2b	$\rightarrow$ 2a	1d	$\rightarrow$	1c
2.	2a	$\rightarrow$ 3b	1c	$\rightarrow$	2d
3.	3b	$\rightarrow$ 1b	2d	$\rightarrow$	3c
4.	1b	$\rightarrow$ 2a	3c	$\rightarrow$	3d

(92) *Problem.*   The animals went in two by two,
But some came out by scores.
For little ones born inside the Ark
Came also through its doors.

The rabbits, the rats, the mice, and cats
Had multiplied indeed.
Cooped up in the hulk for ten long months,
What else to do but breed?

For every two beasts that first went in,
There came out twenty-three:
An average, of course, for all the pairs
In that menagerie.

But some of the beasts, the elephants
And other mammals too,
Came out of the Ark as they'd gone in:
Still two by two by two.

The rest of the horde, a seething mob,
Three hundred pairs all told,
Had bred in the Ark and so came forth
Increased just fifteen-fold.

Imagine that scene on Ararat
For Noah and all his kin,
As the animals fanned across the land!
How many beasts went in?

Solution. Let x be the number of non-breeders. There were 600 breeders which entered the Ark. The number of non-breeders leaving the Ark was again x, whereas the number of breeders increased 15-fold to 9000. Overall the increase in population was by a factor of 11.5, yielding two representations for the population leaving the Ark:

$$11.5(x + 600) = x + 9000.$$

When we solve this for x, we obtain $x = 200$. Thus, the total number of animals entering the Ark was $200 + 600 = 800$.

(93) *Problem.* The three children trooped in, hot and excited. "We had a race," Jack told his mother. "All the way from Main Street." "Who won?" she asked.

"Guess who," replied the boy. "Ann takes twenty-eight steps to run as far as Doug does in twenty-four steps or I do in twenty-one. But I take six steps while Doug takes seven and Ann eight." Well! Who won?

Solution. Let d be the distance for 28 of Ann's steps, 24 of Doug's and 21 of Jack's. Then the length of their respective steps is $d/28$, $d/24$ and $d/21$. Let t be the time required for Ann to take 8 steps, Doug to run 7 and Jack to run 6. Thus, the rates for the three are 8, 7 and 6 steps per unit of time, respectively. Therefore, the three rates of travel are $8(d/28)$, $7(d/24)$ and $6(d/21)$. We need only determine which of the fractions $8/28$, $7/24$, $6/21$ is largest. Clearly, $7/24$ is the largest and the winner of the race is Doug.

(94) *Problem.* Christmas 1984 was over and the three teenagers were looking forward to their birthdays. Jason was born in February, while Mike and John have their birthdays the previous month. John is just eight weeks older than Jason, but three days younger than Mike. What year were they born?

Solution. February 28 is exactly 58 days after January 1. Since there are 59 days between Mike's and Jason's birthdays, it is clear that Mike's birthday is January 1 and Jason's is February 29, making the year they were born as a leap year. Since they are all teenagers and leap years are divisible by 4, they were born in 1968.

(95) *Problem.* Suppose that each of n people knows exactly one piece of information, and all n pieces are different. Every time person A phones person B, A tells B everything he knows, while B tells A nothing. What is the minimum number of phone calls between pairs of people needed for everyone to know everything?

Solution. Consider first the following strategy: Have everyone call person A, say, and pass on his piece of information; this results in $(n-1)$ phone calls; having gathered all the information person A now calls the others back and fills them in on everything, resulting in another $(n-1)$ phone calls for a total of $2(n-1)$ phone calls. This demonstrates that at most $2(n-1)$ phone calls are needed. On the other hand person A needs to receive information from each of $(n-1)$ other people, which requires at least $(n-1)$ phone calls.

In addition he needs to give his information to each of the others, which results in at least $(n-1)$ different calls. Thus, the minimum number of phone calls required is $2(n-1)$.

Thus, $2(n-1)$ phone calls is the minimum number of calls and we have a strategy described which works for that number.

(96) *Problem.* Three teams, A, B and C, compete in a track meet with 10 events. Three points and a gold medal are awarded for first, 2 points and a silver medal are awarded for second, and 1 point and a bronze medal are awarded for third. Team C wins more gold medals than either team A or B. Also team C wins one more medal than team B and two more than team A. Nevertheless, team A comes in first with one point more than team B and two points more than team C. Determine the number of medals of each type won by each team.

Solution. Observe first that there are six points and three medals awarded in each of the ten events for a total of 60 points and 30 medals. From the information supplied, it can be seen that team A was awarded 9 medals and 21 points, team B was awarded 10 medals and 20 points and team C was awarded 11 medals and 19 points. Since team C wins more gold than either of teams A or B, team C must have won at least 4 gold medals, which would account for 12 of its 19 points. Since team C needs 7 more medals and also 7 points, it is clear that the rest of its medals must be bronze. Thus, team C has 4 gold, 0 silver and 7 bronze.

On the other hand team A has at most 3 gold medals since it must have less than team C. This accounts for only 9 of its 21 points. Team A then must have another 12 points with 6 medals, none of which are gold. The only way this can happen is if all 6 medals are silver. Thus, team A has 3 gold, 6 silver and 0 bronze.

This leaves team B with the remainder, namely 3 gold, 4 silver and 3 bronze.

(97) *Problem.* A coin game requires:

1. Nine coins in one pile.
2. That each player take one, three, or four coins from the pile at alternate turns.
3. That the player who takes the last coin wins.

When Adam and Bill play, Adam always goes first and Bill goes second. Each player always makes a move that allows him to win,

if possible; if there is no such move, then he always makes a move
that allows him to tie, if possible.

Must one of the two men consistently win? If so, which one?

Solution. Let us determine whether the player faces a winning
position or a losing position for each possible number of coins remaining when he is about to take his turn. Clearly, a successful
strategy is to leave the opponent with a losing position after your
play. If you can do this, then you are facing a winning position.

# of coins	win or lose	reasons
1	w	take last coin
2	l	can only take 1 and leave 1
3	w	take all 3
4	w	take all 4
5	w	take 3, leave losing position of 2
6	w	take 4, leave losing position of 2
7	l	must leave a winning position
8	w	take 1, leave losing position of 7
9	l	must leave a winning position, since 7 cannot be reached

Since Adam plays first and faces a losing position, Bill must win.

(98) *Problem.* Four men were playing a card game in which (a) a
player must play a card in the suit led, if possible, at each trick
(otherwise, a player may play any card); (b) a player who wins a
trick must lead at the next trick. Nine tricks had already been
played and there were four more tricks left to be played.

1. The distribution of suits in the four hands was as follows:
   HAND I: club – diamond – spade – spade
   HAND II: club – diamond – heart – heart
   HAND III: club – heart – diamond – diamond
   HAND IV: club – heart – spade – spade
2. Art led a diamond at one trick.
3. Bob led a heart at one trick.
4. Cal led a club at one trick.
5. Dan led a spade at one trick.
6. A "trump" won each trick. (A trump is any card in a
   certain suit that may be (a) played when a player has no
   cards in the suit led – in this event a card in the trump suit

beats all cards in the other three suits; or (b) led, as any other suit may be led.)

7. Art and Cal, who were partners, won two tricks; Bob and Dan, who were partners, won two tricks.

Which one of the four men won the fourth trick?

Solution. Since all four players led on the last four tricks, we can conclude that all of these tricks, except possibly the last one, were won by different people. Since a trump won every trick, and since at least three different people won a trick, it is clear that spades is not trump.

Since there are four cards in each suit distributed among the four hands, there are exactly four trump cards at the table, one of which gets played at each trick. Since all four suits are led on one of the tricks, the trump must be led on the last trick.

Thus the person who leads to the last trick must have two trumps in his hand (one to lead to the last trick and another in order to win the third trick which gave him the right to lead to the last trick). This argument shows that clubs cannot be trump.

We now have two possibilities for trump: hearts or diamonds. Since each partnership must capture two tricks, the partner of the person with two trumps must have none, and thus must lead to the first of these last four tricks. If diamonds are trump, we see that Art has two trumps, and his partner Cal must lead to the first trick.

However, Cal must lead a club; in this case everyone must follow suit, and no one can play a trump. This is clearly impossible. Therefore, hearts is trump, Bob has two trumps and his partner, Dan, has no hearts. Bob leads a heart on the last hand and wins the last trick.

(99) *Problem.* Hubert was one of a group of men hired by a jewelry company as an early-morning watchman.

> 1. For no more than 100 days Hubert was on a rotating system of standing watch.
>
> 2. Hubert's first and last watches were the only ones of his to occur on a Sunday.
>
> 3. Hubert's first and last watches occurred on the same date of different months.
>
> 4. The months in which Hubert's first and last watches occurred had the same number of days.

In which one of the twelve months did Hubert have his first watch?
Solution. From clues 2 and 3, we can conclude that the first and
last months of his watch contained the same number of days and
that the difference in days between them is a multiple of 7. This
clearly eliminates the possibility of back-to-back months and also
the possibility of having one intervening month. Consequently,
there are two intervening months, since his watch is no longer than
100 days.

In order for the first and last months of his watch to have the same
number of days we have only 3 possibilities for the month of his
first watch: May, July and December. Of these, the only one which
permits the difference in days to be a multiple of 7 is December,
and then only when the intervening February has 29 days.

Thus, Hubert had his first watch in December preceding a leap
year.

(100) *Problem.* Milly Miffin made a muffin more than Molly's mother
made and Milly Miffin's mother made a muffin more than Molly
made; and Milly, Molly's mother, Molly, and Milly's mother made
fifty muffins, but Milly and Molly's mother made four muffins more
than Molly and Milly's mother made. How many muffins did Milly
make?

Solution. Let x be the number of muffins that Milly made and let
y be the number of muffins that Milly's mother made. Then Molly
made $y - 1$ muffins and Molly's mother made $x - 1$ muffins. In
total they made 50 muffins:

$$\text{that is,} \quad x + (x - 1) + y + (y - 1) = 50$$

$$\text{or} \quad 2x + 2y = 52,$$

$$\text{whence,} \quad x + y = 26.$$

On the other hand Milly and Molly's mother made 4 more muffins
than Molly and Milly's mother, which means

$$x + (x - 1) = y + (y - 1) + 4,$$

$$2x = 2y + 4,$$

$$\text{or} \quad x = y + 2.$$

These two equations can be solved to get $x = 14$ and $y = 12$.
Therefore, Milly Miffin made 14 muffins.

(101) *Problem.* The three bears wake up after a long winter's sleep.
They want to fill their swimming pool with water for a visit from

Goldilocks. They carry water in their bowls. Father bear always works slowly with his big bowl, and he can fill the pool in four hours. Mother bear always works quickly with her medium bowl and she can fill the pool in three hours. Baby bear always does his best, but his tiny bowl means he takes nine hours to fill the pool all by himself. If all three work steadily at the job, how long will it take to fill the pool?

Solution. Let the pool contain V cubic units of water. Then father bear fills the pool at the rate of $\frac{1}{4}V$ cubic units per hour, mother bear fills the pool at the rate of $\frac{1}{3}V$ cubic units per hour and baby bear fills the pool at the rate of $\frac{1}{9}V$ cubic units per hour. Therefore, all three bears working together fill the pool at the rate of

$$\frac{1}{4}V + \frac{1}{3}V + \frac{1}{9}V = \frac{25}{36}V.$$

Thus, the total time required to fill the pool is the volume divided by the rate, implying that it takes $36/25$ hours or 1 hour 26 minutes and 24 seconds.

(102) *Problem.* Five candidates ran in an election. A prominent pollster predicted they would finish in the order $ABCDE$. This was a very bad prediction. Not only did no candidate finish in the place predicted but no two candidates predicted to finish consecutively actually did so in the order predicted. A second prediction had the candidates finishing in the order $DAECB$. This prediction was better. Exactly two of the candidates finished in the places predicted, and two disjoint pairs of candidates predicted to finish consecutively actually did so in the order predicted. Determine the actual order of finish.

Solution. The second prediction indicates that there are two disjoint pairs of candidates that finished consecutively in the order predicted. Note that if one member of such a pair is in the correct position, then so is the other. This implies that exactly one of these pairs is not only in the correct order but in the correct position. It can be determined that this pair must be at either end, that is, either DA or CB. Thus, there are only two choices for the actual order by considering the second prediction: $DACBE$ or $EDACB$. However, $DACBE$ matches prediction for candidate C, which is impossible. Therefore, the actual order of finish is $EDACB$.

(103) *Problem.* Paul looked down at the shiny new license plates on the table. The three friends had gone together to renew their car licenses and were now having some refreshment. "It's all nonsense having new plates every year," he remarked. "Just as I begin to remember my number it has to be changed."

"That goes for me, too," agreed Dick, "but there's something funny I've noticed about our numbers this time. Your first figure is the same as Hal's last, and your last the same as his first, and you've both got the same figures in the middle."

"And there's something else," exclaimed Hal. "Your two numbers add up to mine."

Paul picked up the check which his friends seemed too engrossed to have noticed. "Let's go," he said, "but I would have you note that Dick's two middle figures are the same and are also the same as your first figure, and his first figure is the same as mine."

Well, it certainly was a mix-up of figures among them; fortunately, all three had been given 4-figure numbers. So, now what was Dick's license number?

Solution. From the above information, we may conclude that Paul's license number has the form $ABBC$, Dick's has the form $ACCD$, and Hal's has the form $CBBA$. In addition, we are told that $ABBC + ACCD = CBBA$. Since the 2 middle digits BB stay BB after the addition, we may conclude that $C = 0$ or 9. But since the leading digits add up to C, we must conclude that $C = 9$. This in turn implies that $A = 4$ and $D = 5$. Therefore, Dick's license number is 4995. It is interesting to note that we have no information on the value of B!

(104) *Problem.* "Let's not play for point-stakes today," said Mary as the three friends sat down to play Canasta. "I suggest the loser should double whatever money each of the other two has at the end of the game." Her two guests agreed with this novel idea – after all, Mary was hostess – and so they settled down to serious business. Gwen was down on her luck at first: she barely managed to meld, and she lost the first game by a disgraceful margin. She made up for it in the next game, however, which she won.

Mary lost the second game. But the third was really exciting. It developed into a hard tussle between Mary and Gwen, each drawing lots of Wild Cards and going out several times on Concealed Hands. After that third game, in which Clare had hardly scored any points at all, Mary proposed a break for tea. Pouring for her guests some minutes later, she remarked that she had won a little. "I've got exactly sixteen dollars now," she said. The other two checked their cash also and found they each had sixteen dollars too.

It was a strange coincidence. How much did Gwen have when they started to play?

Solution. Let g, m, and c be the amounts of money that each of Gwen, Mary, and Clare had when they started to play. After game #1:

$$\begin{aligned} \text{Gwen had } & g - m - c, \\ \text{Mary had } & 2m, \\ \text{Clare had } & 2c, \end{aligned}$$

and after game #2, which Mary lost, we have

$$\begin{aligned} \text{Gwen at } & 2(g - m - c), \\ \text{Mary at } 2m - (g - m - c) - 2c = & 3m - c - g, \\ \text{Clare at } & 4c. \end{aligned}$$

Since Clare lost the third game, the dollar amounts are adjusted to:

$$\begin{aligned} \text{Gwen } & 4(g - m - c) = 16, \\ \text{Mary } & 2(3m - c - g) = 16, \\ \text{Clare } 4c - (2g - 2m - 2c) - (3m - c - g) = 16, \end{aligned}$$

which resolves into the system:

$$\begin{aligned} 4g - 4m - 4c &= 16, \\ -2g + 6m - 2c &= 16, \\ -g - m + 7c &= 16. \end{aligned}$$

This system has solution $g = 26$, $m = 14$, and $c = 8$, whence Gwen started with $26.

(105) *Problem.* At a party Jack saw Jill standing alone at the punch bowl.

1. There were nineteen people altogether at the party.

2. Each of seven people came alone; each of the rest came with a member of the opposite sex.

3. The couples who came to the party were either engaged to each other or married to each other.

4. The women who came alone were unattached.

5. No man who came alone was engaged.

6. The number of engaged men present equalled the number of married men present.

7. The number of married men who came alone equalled the number of unattached men who came alone.

8. Of the married women, engaged women, and unattached women present, Jill belonged to the largest group.

9. Jack, who was unattached, wanted to know which group of women Jill belonged to.

Which one of the three groups of women did Jill belong to?

Solution. Let C be the number of married couples present, A the number of men who came alone, and B the number of unattached men. We can express many of the above conditions in the following tree:

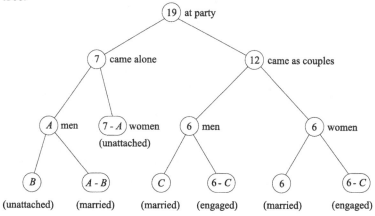

To summarize: there are $7 - A$ unattached women, C married women, and $6 - C$ engaged women. We also have

$$6 - C = (A - B) + C \qquad \text{from statement \#6}$$
$$\text{and} \qquad A - B = B \qquad \text{from statement \#7.}$$

This allows us to deduce that $B = A/2$ and that $6 - C = A/2 + C$, whence $A/2$ must be an even integer, and thus, A itself must be a multiple of 4. Since A is at most 7, it must be either 0 or 4. But Jack is unattached; this implies that $B > 0$ and therefore $A > 0$. So $A = 4$, $C = 2$, $B = 2$. So there are 3 unattached women, 2 married women and 4 engaged women, which implies that Jill is engaged!

(106) *Problem.* Today I met a monkey, but this was downtown where they are building that new store. He was dangling at the end of a rope, holding on with one paw and scratching happily with the other. Going closer, I saw that the rope passed over a pulley with

a weight at the other end, and the two ends were exactly level. This was all so strange that I made some inquiries from a man who seemed to be working there. Thus I learned that the monkey's age and the age of its mother total seven years; furthermore, the monkey weighs as many pounds as its mother is years old.

Intrigued by these revelations, I heard that the monkey's mother is one third again as old as the monkey would be if the monkey's mother were half as old as the monkey will be when the monkey is three times as old as the monkey's mother was when the monkey's mother was three times as old as the monkey was then.

This news staggered me, but more was to come: the weight of the weight and the rope came to twice the difference between the weight of the weight and the weight of the weight and the monkey. I could take no more. "Tell me the length of the rope," I pleaded, "and let me go." My informant shook his head sadly: "That we don't know," he said, "but four feet of it weigh one pound."

I went on my way wondering "How old is the monkey and how long is the rope?"

Solution. Let the monkey's age be x and let the monkey's mother's age be y. Then clearly $x + y = 7$ and also the monkey weighs y pounds. Since the monkey and the weight on the rope are in equilibrium, we can conclude that the values are the same. From paragraph #3 we may also deduce that the monkey and the rope weigh the same amount. That is, the rope weighs y pounds. Since the rope weighs $\frac{1}{4}$ pound per foot, the length of the rope is $4y$ feet.

Let t be the time elapsed since the monkey's mother was 3 times as old as the monkey was. When the monkey's mother was 3 times as old as the monkey was then, the monkey's mother was $y - t = 3(x - t)$ years old. When the monkey was 3 times as old as this, the monkey was $9(x - t)$ years old. If the monkey's mother were half as old as this, she would be $4.5(x - t)$ years old, and the monkey would be $y - x$ years younger or $5.5x - 4.5t - y$ years old. But the monkey's mother is one third again as old as this, that is,

$$y = \tfrac{4}{3}(5.5x - 4.5t - y),$$

from which we see that $7y = 22x - 18t$.

But t years ago we have $y - t = 3(x - t)$. Eliminating t from these two equations yields $2y = 5x$. This together with the equation

$x + y = 7$ gives the solution $(x, y) = (2, 5)$. Thus, the monkey is 2 years old, its mother is 5 years old, and the rope is $4y = 20$ feet in length.

(107) *Problem.*   This elegant puzzle dates back to at least 1739. For historical interest I give it in the original dress, which seems to have imposed the English currency on the Netherlands. I hasten to add that all you need to know about this currency is that a guinea contains 21 shillings.

Three Dutchmen and their wives go to market, and each individual buys some hogs. Each buys as many hogs as he or she pays in shillings for one hog. Each husband spends altogether 3 guineas more than his wife. The men are named Hendrick, Elas, and Cornelius; the women are Gurtrun, Katrun, and Anna. Hendrick buys 23 more hogs than Katrun, while Elas buys 11 more than Gurtrun. What is the name of each man's wife?

Solution.   The amount spent by each person is a perfect square, and the difference of the expenditures within each family is 63 shillings. First we determine how many pairs of squares differ by 63:

$$h^2 - w^2 = 63$$
$$(h - w)(h + w) = 1 \cdot 3 \cdot 3 \cdot 7$$

where h represents the number of hogs bought by the husband and w the number bought by the wife from a particular family. Since $h - w$ is smaller than $h + w$ and is an integer, it must be one of 1, 3, or 7, which leads to the solutions $(h, w) = (32, 31)$, $(12, 9)$, and $(8, 1)$. Since 2 differences are given we see that the number of hogs bought by each was:

Gurtrun 1   Elas 12

Katrun 9   Hendrick 32

which further implies that Anna bought 31 and Cornelius 8. This establishes the couples as: Anna and Hendrick, Katrun and Elas, Gurtrun and Cornelius.

(108) *Problem.*   The ingenious manner in which a box of treasure, consisting principally of jewels and precious stones, was stolen from Gloomhurst Castle has been handed down as a tradition in the De Gourney family.

The thieves consisted of a man, a youth, and a small boy, whose only mode of escape with the box of treasure was by means of a high window.

Outside the window was fixed a pulley, over which ran a rope with a basket at each end. When one basket was on the ground the other was at the window.

The rope was so disposed that the persons in the basket could neither help themselves by means of it nor receive help from others. In short, the only way the baskets could be used was by placing a heavier weight in one than the other.

Now, the man weighed 195 lbs., the youth 105 lbs., the boy 90 lbs., and the box of treasure 75 lbs.

The weight in the descending basket could not exceed that in the other by more than 15 lb. without causing a descent so rapid as to be most dangerous to a human being, though it would not injure the stolen property.

Only two persons, or one person and the treasure, could be placed in the same basket at one time. How did they all manage to escape and take the box of treasure with them?

Solution. We will try to show not only how to allow all to escape, but to show all possible ways that they can escape.

First let us show that each stable position (that is, the baskets are not moving) can be characterized by an ordered pair describing the set at the bottom and the set at the top.

For example, $(-, tbym)$ would represent the initial position with nothing at the bottom, and (by, tm) would represent the position with the boy and youth at the bottom and the man and the treasure at the top.

To solve, we need to consider all 16 possible combinations as points and then draw an arrow between two points if in one move of the baskets we can change from one combination to the other.

We then seek a path of arrows from $(-, tbym)$ to $(tbym, -)$. Upon proper organizing of the 16 combinations, the graphical solutions would be as shown:

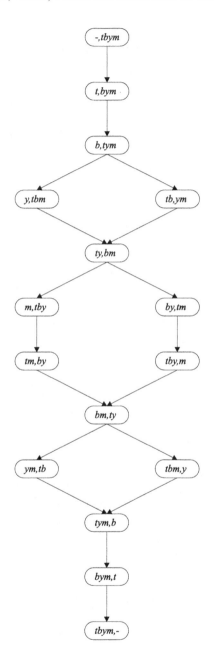

(109) *Problem.*   Mary, Karl, John, Andy and Peter went looking for
mushrooms. Only Mary took her search seriously. When it was
time to return, Mary had 45 mushrooms and the boys had none.
Mary was sympathetic. "It won't look good for you boys when we
get back to camp." She gave each boy some mushrooms, leaving
none for herself.

On the way back Karl found 2 mushrooms and John doubled the
number of mushrooms he already had. But Andy and Peter fooled
around all the way. As a result Andy lost 2 mushrooms and Peter
lost half his mushrooms.

At camp they counted up and discovered that each boy had the
same number of mushrooms. How many mushrooms did Mary give
each boy?

Solution.   Let x be the number of mushrooms that each boy had
when he returned to camp. Originally, Karl was given $x - 2$ mushrooms by Mary, John was given $x/2$ mushrooms, Andy was given
$x + 2$, and Peter was given $2x$ mushrooms. Since Mary divided all
her 45 mushrooms among the 4 boys we get

$$(x - 2) + (x/2) + (x + 2) + (2x) = 45$$
$$\text{or} \quad 4.5x = 45$$

which yields $x = 10$. Therefore, originally Karl was given 8 mushrooms, John 5, Andy 12 and Peter 20.

(110) *Problem.*   A man bought an odd lot of wine in barrels and one
barrel containing beer. The number of gallons contained in these
6 barrels were (in no particular order) 15, 31, 19, 20, 16, and 18.
He sold a quantity of the wine to one man and twice the quantity
to another, but kept the beer to himself. The puzzle is to determine the number of gallons there were in the barrel of beer. Of
course, the man sold the barrels just as he bought them, without
manipulating in any way the contents.

Solution.   Let x be the quantity of wine sold to the first man. Then
$2x$ is the amount sold to the second man. Let y be the amount of
beer. Then

$$x + 2x + y = 15 + 31 + 19 + 20 + 16 + 18 = 119.$$

This can be rearranged to yield

$$y = 3(40 - x) - 1,$$

which means that the amount of beer, y, is 1 less than a multiple of 3. The only value that works is $y = 20$. Therefore, there are 99 gallons of wine from which the first man bought 33 gallons and the second bought 66. The only way to get 33 gallons is by combining the 15 and 18 gallon kegs. Thus, the entire solution is:

$$x = 15 + 18 = 33,$$
$$2x = 31 + 19 + 16 = 66,$$
$$y = 20.$$

(111) *Problem.* In some faraway land there are 3 types of people named Lats, Alts, and Tals, who always lie, alternate in lying and telling the truth, and always tell the truth, respectively.

Three residents each made 2 statements, as follows:

1st man: Both of you are Tals.
2nd man: Both of you are Alts.
3rd man: One of you is a Tal and the other is an Alt.

1st man: I am a Tal.
2nd man: I am a Tal.
3rd man: I am a Tal.

To which type does each man belong?

Solution. By considering the first two statements, we see that the first statement must be false. Since this makes the fourth statement also false, we may conclude that the first man is a Lat. Since the first man is a Lat, the second man's first statement is false, a fact which makes his second statement also false, whence the second man is also a Lat. Since both the first man and second man are Lats, the third man's first statement is now false, a fact which implies that his second statement is also false, and thus, he also must be a Lat.

Therefore, all three men are Lats, that is; they always lie.

(112) *Problem.* During the summer of 1983, the MacLeonard's Corporation ran a promotional game. With each purchase a customer received a game card. Each game card had ten spaces covered from view. In eight of these spaces, the names of eight different products appeared. In a ninth space the name of one of the eight products appeared again.

Thus, these nine spaces contained only one possible pairing of product names. The remaining space contained something different; we

shall call it an "X". To play the game card spaces were uncovered one at a time in any order. Play ceased when either the pair
had been uncovered (a "win" of that product) or the X appeared
(a loss). Thus, every card could be a "winner". What was the
probability of winning this game?

Solution. For any given card there are only three symbols that
have any bearing on the probability, namely the repeated symbol
and the X symbol. Among these three symbols, there are three
equally probable events: the X could be scratched first, second, or
not scratched (that is; left till last). Only the last of these three
events represents a win. Thus, the probability of winning is 1/3.

(113) *Problem.*   A woodsman paddling steadily across the still surface
of a northern lake saw a magnificent bass break water directly ahead
of him. Twelve strokes he counted until his canoe first crossed the
ever-widening circle the fish had made, and then twelve more before
he broke through the circle on the opposite side. How far away was
the bass from the woodsman at the moment it jumped?

Solution. In this problem all distances are measured in strokes.
Let x be the distance in strokes of the bass from the woodsman at
the moment it jumped. Then when he first crossed the wave made
by the fish, he was still $x - 12$ strokes from the spot where the fish
jumped. Therefore, while the woodsman paddles 12 strokes, the
wave moves $x - 12$ strokes. Thus, during the next 12 strokes that
the woodsman makes the wave moves out a further $x - 12$ strokes.
Therefore, the diameter of the small circle he first encounters is
$2(x - 12)$ strokes, and then he must travel yet another $x - 12$
strokes to overtake the enlarging circle, and during this time he
has moved another 12 strokes. So we get

$$3(x - 12) = 12,$$

from which we can conclude that x is 16 strokes.

(114) *Problem.*   Three people are playing table tennis, and after every
match the loser gives up his place to the person not playing. At
the end the first player has played 10 games and the second has
played 21. How many games has the third player played?

Solution. Let n be the number of games played between the first
and second players. Then n is at most 10, since the first player only
played 10 games. Now $21 - n$ games have been played between the
second and third players, and in between any two such games (in

$21 - n - 1$ instances) there must have occurred at least one game played by the first player. Therefore,

$$21 - n - 1 \le 10$$
$$\text{or} \quad n \ge 10.$$

Thus, $n = 10$ and it follows that the third player has played 11 games.

(115) *Problem.* What is a polyomino?[1] It is a structure made of unit squares or cells joined along their sides. A single square is called a monomino. Two make a domino. Three join in two different ways to make two triominoes. Four join to make five tetrominoes, as shown below:

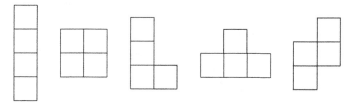

Together the five four-cell structures make up 20 cells. Can you "tile" a 4×5 rectangle with them? By "tiling" we mean that the five structures should completely cover the rectangle without overlapping each other or "overhanging" the rectangle.

Solution. If you checkerboard colour each structure, you will notice that all but one of the structures has the same number of lightly and darkly coloured cells; the exception is the (inverted) T-shaped one, which has three cells of one colour and one of the other colour. In a 4×5 array a checkerboard colouring of the cells has the same number of lightly and darkly coloured cells, and thus could **not** possibly be coloured by the five tetrominoes.

If the T-shaped piece is replaced by a duplicate of any of the other pieces, then the problem can be solved. Also, if any of the other pieces is replaced by a second T-shaped piece, then the problem can again be solved. You might enjoy trying these puzzles.

(116) *Problem.* "All aboard!" called Mary, stepping gingerly into the canoe from the slippery steps at the foot of the garden. She settled herself into the cushions while her husband started to paddle upstream; she yawned, she slept.

---

[1]The name "polyomino" was invented by Solomon Golomb in 1953. See *Polyominoes*, Princeton University Press.

The noise of traffic overhead wakened her as they passed under the bridge some minutes later. She opened her eyes, looked over the side, and screamed: "Steve, there's a hand in the water! It's a dead body!"

"You've been dreaming, darling" he laughed, seeing what she saw. "It's only an old glove full of nothing." Relaxing with a sigh of relief, Mary closed her eyes again, and Steve went on paddling doggedly upstream.

After a further fifteen minutes he turned and made for home. "Whoosh, whoosh, whoosh," went his paddle – always the same steady rhythm as the light shell slipped through the water. They shot under the bridge, and then it didn't take long to make the remaining mile to their house.

As they arrived abreast their steps, Steve roused his wife with a shout: "Wakey, wakey! Here we are, and there's your corpse." And there indeed, still floating in mid-stream, was that very lifelike old glove which they had just overtaken.

Steve paddled at the same speed all the time. But what do you make the speed of the current?

Solution. This is one of those problems where there seems to be insufficient information to solve the problem. However, that is only an illusion.

Let x be the speed of the boat on the water and let y be the speed of the water with respect to the land. Both speeds are assumed to be measured in miles per hour. Also let d be the distance from the bridge to turn-around point. Since Steve paddled for $\frac{1}{4}$ hour upstream at a speed relative to the shore of $x - y$ miles per hour, we see that $d = (x - y)\frac{1}{4}$. The time required for the glove to travel from the bridge to their house is $1/y$ hours. During this time Steve paddles upstream for $\frac{1}{4}$ hour and then paddles $d + 1$ miles downstream at a speed of $x + y$ miles per hour relative to the shore. Thus,

$$\frac{1}{y} = \frac{1}{4} + \frac{d+1}{x+y}.$$

Therefore,

$$\frac{x-y}{4} + \frac{1}{x+y} + \frac{1}{4} = \frac{1}{y},$$

$$x - y + 4 + x + y = \frac{4x + 4y}{y},$$

$$2xy + 4y = 4x + 4y,$$
$$\text{or} \qquad x = 2,$$

whence the velocity of the current is 2 miles per hour.

(117) *Problem.*   A number of years ago, I accompanied Indiana Jones to his last archaeological dig: the vault of the great Pharaoh of Laffa. The joy of seeing first hand this tremendous discovery was soon replaced by such despair, since I hope no one should ever have to experience, for the stone over the entrance, having held steady for thousands of years, picked the very moment of our entry to fall and seal us in!

"Not luck," said Jones coolly.   "The ancient engineers booby-trapped the vault.   Maybe they rigged an escape route.   Shine the light on the door."

I did so, and Jones read: "Cursed be ye, if trespassers! Otherwise, press the two keystones to open the door. Yet beware that exactly four stones are liars."

"The keystones," said Jones, "must be two of the stones projecting from the walls. There's one at the centre of each wall — north, south, etc. — and one in each corner — northwest, southwest, etc."

Jones then translated the inscriptions on the stones:

(a) NW: Exactly one keystone is in a corner.

(b) N: At least one keystone is a truthteller.

(c) NE: The keystones are side-by-side.

(d) E: No three lying stones are side-by-side.

(e) SE: At least one keystone is a liar.

(f) S: Exactly two corner stones are liars.

(g) SW: The longest unbroken 'row' of liars is three stones.

(h) W: The keystones are both liars, or both truthtellers.

"Ingenious," said Jones. "You'll note that the ceiling is of the same stone as the walls, but not joined to them. Doubtless pressing the wrong ones would bring it crashing down. But one must press two at once ..."

"Not to change the subject, but is it getting stuffy in here?"

"Hmm, can't tell that it is. Well then, if you'll take that stone, I shall press this one, and we will be out of here in a jiffy."

Which stones were the liars and which were the truthtellers? More importantly, which stones were the keystones?

Solution. First consider the stones N, SE and W. Observe that exactly one them is false, and two true. Next consider the stones E and SW. One of these must be true and one must be false. This implies that of the remaining stones, that is, NW, NE, and S, exactly one is true and two are false. Among these last three stones we can also not that if NE is true, then NW is automatically true, which is impossible since at most one them can be true. Thus we may conclude that NE must be false.

Since NE is false, one of stones NW or S is false and one is true. Let us first assume that S is true and NW is false. Then stone S says that SE and SW are both true, whence E is false. But with this setup we cannot assign true/false values to the remaining stones to get 3 liars in a row, which means that SW must be false, a contradiction.

Thus NE is false, S is false and NW is true. Since stone S is false, we can deduce that SW and SE are both true or both false. If they are both false, then we have determined our 4 false stones and the remaining stones, namely E, N, and W are true. But this very arrangement of true/false values contradicts E. Therefore, we must conclude that SW and SE are both true. Since SW is true, we have E is false. From this we can determine that N is false and W is true.

So the liars are NE, S, E and N; and the truthtellers are NW, SW, SE and W. Since N is false, no keystone is a truthteller. From this and from NW we see that NE is a keystone. Stone W now tells us that both keystones are liars. Stone NE false implies that the keystones are not side-by-side. Since the only liar which is not beside NE is S, we conclude that S is the other keystone. Therefore, the keystones are NE and S.

(118) *Problem.* Susan cut across the middle of a chess board, splitting it in half, and wrote a digit in each square. The top eight-digit number was factorable by 3, the second by 5, the third by 7, and the bottom number by 11. In addition the numbers alternated between even and odd. She left the room for several minutes. When she returned, she found that her younger brother Winston had cut out each of the squares. He had kept all eight digits of each number together, but the digits were no longer in their proper order, nor were the numbers themselves in the original order. The squares looked like this:

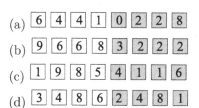

(a) 6 4 4 1 0 2 2 8

(b) 9 6 6 8 3 2 2 2

(c) 1 9 8 5 4 1 1 6

(d) 3 4 8 6 2 4 8 1

In what order were the rows originally arranged, and what was the number in the original third row (the number factorable by 7)?

Solution. For numbers which are factorable by 11, alternately adding and subtracting the digits also results in a number which is factorable by 11. The colours of the squares identify the alternate digits, from which (c) is seen to be the only number factorable by 11, no matter what the order. Numbers (a) and (c) are the only ones containing a 5 or a 0. Since (c) is already identified as the number factorable by 11, (a) must be the number factorable by 5. From the sum of the digits, (a) and (d) are factorable by 3. Since (a) is already identified as the number factorable by 5, (d) must be the top number.

Thus, the order of the numbers is (d), (a), (b), (c).

One can exhaustively check all the numbers in made from digits in row (b) alternating between white and black squares to see which are factorable by 7; this yields only the following set:

$$26382629$$
$$29382626$$
$$62628293$$
$$62926283$$
$$62928362$$
$$63628292$$
$$63926282$$
$$82926263$$
$$83926262$$

Since (a), which immediately preceded (b) in the original list, ends in a black square, 0, then (b) must end in a white square. Only the first two in the above list end in a white square. Also (a), which ends in a 0, is an even number. Therefore, (b) must be odd, whence (b) is 26382629.

(119) *Problem.*   A man has a bowl that holds a little more than a pint, and a flat rectangular straight-sided pan that holds exactly a pint. He wants to put exactly one-third of a pint of water into the bowl, but he has no other means of measuring anything. He has a supply of water and an ordinary kitchen table with an exactly level surface. How does he do it?

Solution. He fills the pan on the table more than half-full, and then carefully tilts one side, pouring out the water, until the level reaches the bottom edge of the raised side. This leaves exactly one-half pint in the pan, since the empty part is the same shape and size as the filled part.

Being an ordinary kitchen type the table has a straight edge; the man slides the pan over the edge so that the opposite corners of the pan coincide with the edge of the table, and starts tilting again, this time catching the water with the bowl. He tilts until the surface of the water coincides with the corners of the pan. The bowl now contains one-third pint.

Suppose that the length, width and height of the pan are l, w and h, respectively. At the point when the water surface coincides with the corners of the pan, the water forms a pyramid, the volume of which is one-third the area of the base times the height. Here the base is $\frac{1}{2}lw$, which implies that the volume is

$$\frac{lw}{2} \times \frac{h}{3} = \frac{lwh}{6}.$$

Since the volume of the pan is lwh (1 pint), the remaining amount is one-sixth pint. Thus, he poured $\frac{1}{2} - \frac{1}{6}$ pint, that is, $\frac{1}{3}$ pint.

(120) *Problem.*   The country of Marr is inhabited by two types of people, liars and truars (truth tellers). Liars always lie and truars always tell the truth. As the newly appointed Canadian ambassador to Marr, you have been invited to a local cocktail party.

While consuming some of the native spirits, you are engaged in conversation with three of Marr's most prominent citizens: Joan Landill, Shawn Farrar, and Peter Grant. At one point in the conversation Joan remarks that Shawn and Peter are both liars. Shawn vehemently denies that he is a liar, but Peter replies that Shawn is indeed a liar. From this information can you determine how many of the three are liars and how many are truars?

Solution. First of all note that if two inhabitants disagree on anything, one is a truar and one is a liar. On the other hand if they agree, they must both be the same: either both truars or both liars. Since Shawn and Peter disagree on whether Shawn is a liar, one of them is a liar and one is a truar. Thus Joan's statement is false and she is a liar.

Thus, there are 2 liars and 1 truar among the three of them. From the information given it is not possible to decide which of Shawn and Peter is a truar and which a liar!

(121) *Problem.* Eleven lines are to be drawn in the plane. Exactly three are to pass through point A, exactly three through a different point B, and no other three are to be concurrent.

Determine the <u>minimum</u> number of distinct intersection points, including points A and B.

Solution. Let the first line pass through A and B. Now position the remaining lines through A and B so that they are parallel in pairs. In this way the first 5 lines have been placed so as the create only 4 distinct points of intersection.

Now select one of the lines already placed different from the line through A and B. Place all remaining 6 lines parallel to this selected line. They are clearly parallel to each other and each has only 3 distinct points of intersection with the previously placed lines. This makes for a total of $4 + (6 \times 3) = 22$ distinct points of intersection.

(122) *Problem.* Five county jails—including the one in which Deputy Bennett works—for the first time in a long while are not filled to capacity with miscreants. None of the county jails is large (two of them have only six cells), but by housing two prisoners to a cell they are able to double their capacity. From the clues below, can you determine the sheriff and deputy of each county jail, the number of cells in each jail, and the current number of prisoners each jail is holding?

Note: No sheriff shared an initial with either his deputy or the county in which he was sheriff; neither did any deputy share an initial with the county within which he worked.

> 1. The two jails with the most cells (eight) were in Benton County and the county where Adams was sheriff.
>
> 2. Deputy Dixon, who did not work in Canton County, was in charge of ten prisoners, the same number for which the deputy in Anton County was responsible.

3. Sheriff Dawson, whose deputy was not Edwards, had more prisoners than the sheriff of Eaton County, who had nine.

4. Sheriff Casey (not of Denton County) had a jail larger than Sheriff Bailey's (whose deputy was not Allen) but smaller than Sheriff Eames'.

5. The county jail with the fewest cells (half the size of the Benton County jail) was one prisoner short of capacity and had five fewer prisoners than the number Deputy Clark was responsible for.

Solution. Let us first examine Deputy Dixon. Clue 2 shows us that he cannot work in Anton or Canton Counties; it also states that he was in charge of 10 prisoners, which means that he could not work in Eaton County, which had 9 prisoners according to Clue 3. Since it is given that he cannot work in Danton County, he must work in Benton County.

One piece of given data states that there are two 6-cell jails; Clue 1 gives us two 8-cell jails, and combined with Clue 5 the remaining jail must have 4 cells. Again from Clue 5 this smallest jail must have 7 prisoners (one short of capacity), which shows that some jail has 12 prisoners. Combining this with Clues 2 and 3 tells that the jails have (in some order) 7, 9, 10, 10, and 12 prisoners. Since none of the jails is at capacity, the jail with 12 prisoners must be an 8-cell jail. Clue 5 also tells us that Deputy Clark works for the County with 12 prisoners; this rules out Anton and Eaton from Clues 2 and 3; he also cannot work for Canton County (given), and we have Deputy Dixon in Benton County; this leaves only Denton County for Deputy Clark.

Now Clue 4 indicates that Sheriff Bailey belongs to the 4-cell jail (with 7 prisoners), Sheriff Casey belongs to one of the two 6-cell jails, and Sheriff Eames belongs to one of the two 8-cell jails. From Clue 1 Sheriff Adams has an 8-cell jail, but Adams is not from Benton County, which also has an 8-cell jail. Thus, Sheriff Eames must work for Benton County. We now know all about Benton County: Sheriff is Eames, Deputy is Dixon, 8 cells with 10 prisoners.

Since none of the jails is at capacity, the jail with 12 prisoners (in Denton County) must have 8-cells. Since there is only one 8-cell jail undetermined, and from Clue 1, Sheriff Adams is in charge of it, we now know all about Denton County: Sheriff is Adams, Deputy is Clark, 8 cells with 12 prisoners.

Since Sheriff Bailey's jail has only 4 cells, Clues 2 and 3 indicate that he cannot work for either Anton or Eaton Counties. Since we have already the Sheriff for both Benton and Denton Counties, we conclude that Sheriff Bailey works in Canton County. This only leaves Anton County for Sheriff Dawson and Eaton County for Sheriff Casey.

Clearly Deputy Bennett does not work with Sheriff Bailey. Deputies Clark and Dixon are already paired with Sheriffs, and Clue 4 states that Deputy Allen does not work with Sheriff Bailey. This leaves only Deputy Edwards to work with Sheriff Bailey (in Canton County). This in turn shows that Deputy Allen must work in Eaton County, and Deputy Bennett must work in Anton County. The complete answer is given below:

County	Sheriff	Deputy	# of Cells	# of Prisoners
Anton	Dawson	Bennett	6	10
Benton	Eames	Dixon	8	10
Canton	Bailey	Edwards	4	7
Denton	Adams	Clark	8	12
Eaton	Casey	Allen	6	9

(123) *Problem.* Nine foreign journalists meet at a press conference. Each of them speaks at most three different languages, and any two of them can speak a common language. Show that at least five of them speak the same language.

Solution. Let us suppose that no language is spoken by more than 4 journalists. We must show that this impossible. We can now show that under this supposition each of the 9 journalists must speak exactly 3 languages: for if any of them speaks at most 2 languages, then at least one of these 2 languages must be spoken by at least 4 of the remaining 8 journalists, which would mean that 5 people speak the same language, which is impossible by our initial supposition. Now suppose that there is a language with fewer than 3 speakers, that is, suppose there is a language spoken by at most person A and person B. Of the 2 remaining languages spoken by A at least one of them must be spoken by at least 4 of the remaining 7 journalists, again leading to 5 persons speaking a common language, which contradicts our initial supposition. Thus we may conclude that if no language is spoken by more than 4 speakers, we must have every journalist speaking exactly 3 or exactly 4 languages and every language spoken by at least 3 journalists.

Now let x and y be the number of languages with 3 and 4 speakers, respectively. If we count the number of (speaker, language) pairs in two different ways we get

$$3x + 4y = 9 \cdot 3$$

where the left hand side counts the number of speakers for each language and the right side counts the number of languages spoken by each speaker. Since we are looking for non-negative integer solutions for the above equation, we have only 3 possibilities:

$$(x, y) = (9, 0), \qquad (x, y) = (5, 3), \qquad \text{or } (x, y) = (1, 6).$$

Since each pair of journalists can speak a common language, we get

$$\binom{3}{2} x + \binom{4}{2} y \geq \binom{9}{2},$$

where $\binom{n}{2} = \frac{n(n-1)}{2}$ is the number of ways to choose a pair of objects from a set of n objects. This simplifies to $x + 2y \geq 12$. This leaves only the case $(x, y) = (1, 6)$.

Let the seven languages be represented by a, b, c, d, e, f, and g, where language a is the unique one with 3 speakers. Also let the nine journalists be denoted by A, B, C, D, E, F, G, H, and I, with A, B and C the journalists who speak language a.

Without loss of generality we can assume that journalists A, B and C speak languages as follows:

$$
\begin{array}{ll}
A & a, b, c \\
B & a, d, e \\
C & a, f, g
\end{array}
$$

The remaining six journalists when paired with A must either speak language b or language c; thus 3 of them must speak language b and 3 language c. By a similar pairing with B we also conclude that 3 must speak language d and 3 language e. Same again for languages f and g when paired with C. Without loss of generality we may assume that language b is spoken by D, E, and F, while language *c* is spoken by *G*, *H*, and *I*. If language *d* were also spoken by *D*, *E*, and *F*, then language *e* would be spoken by *G*, *H*, and *I*; this would imply that in order for each of *D*, *E*, and *F* to have a language in common with each of *G*, *H*, and *I* all six would have to speak a common language, which is a contradiction.

Thus, we may further assume that language *d* is spoken by *D*, *G*, and *H*, while language *e* is spoken by *E*, *F*, and *I*. Now each of the journalists *E* and *F* must have a language in common with each of *G* and *H*; the only way this can be done is for all four of them to speak a common language, which must be either *f* or *g*, which is also spoken by *C*. This last situation is also a contradiction since there would be 5 speakers for that language. Therefore, we must conclude that there is a language with at least 5 speakers.

(124) *Problem.*   Five couples live in Confusion Condominium. Their lives are made difficult by the fact that one of the five men makes his living as a thief. The other four husbands are a rich man, a poor man, a beggar man, and a doctor. One member of each couple always tells the truth and the other lies. In four of the five couples, it is the wife who always tells the truth and the husband who lies. The only husband who always tells the truth is the poor man, as you might suspect. A further complication is that all ten persons have names that can designate men or women. In statements made below by the ten persons, nobody mentions the name of his/her own spouse.

Bobby	1	Pat is married to Terry.
Ellie	2	Kim is married to the rich man.
Freddie	3	Ellie is married to Ronnie.
Jerry	4	Ellie is married to Willie.
	5	Freddie is the beggar man.
Kim	6	Jerry is not married to Terry.
Lou	7	Pat is not the poor man.
	8	Bobby is married to Jerry.
Pat	9	Freddie is married to Ronnie.
	10	Willie is not the doctor.
Ronnie	11	Jerry is either the rich man, or Jerry is a woman.
	12	Bobby is married to Kim.
Terry	13	Freddie is married to Lou.
	14	Ellie is a woman.
Willie	15	Kim is the thief.
	16	Freddie tells the truth.

Who is the thief?

Solution. First construct a table indicating who is not married to whom:

	Bobby	Ellie	Freddie	Jerry	Kim	Lou	Pat	Ronnie	Terry	Willie
Bobby	X					X	X	X	X	
Ellie		X	X	X	X				X	
Freddie		X	X	X			X	X	X	X
Jerry		X	X	X	X	X		X		X
Kim		X		X	X			X	X	X
Lou	X			X		X	X		X	
Pat	X		X			X	X	X		X
Ronnie	X		X	X	X		X	X		
Terry	X	X	X		X	X			X	
Willie			X	X	X		X			X

The above table is filled in by noting that no one is married to himself or herself and that no one mentions his or her spouse in any of the statements. Now since Freddie is not married to Ronnie, Pat lied with statement 9. Thus Pat's statement 10 must also be a lie. This means that Willie is the doctor, and thus a liar, which further implies that his statements are lies. Therefore, Freddie lies! (and also, Kim is not the thief). Since Freddie lies, we see that Ellie is not married to Ronnie.

Since statement 9 is false (see above), we note that Pat is not the poor man (who always tells the truth). Therefore, Lou's statement 7 is true, which implies that statement 8 is also true. So Bobby is married to Jerry. (At this point it may be useful to start filling in the table further, using an X for an unmarried pair and a check mark, ($\sqrt{}$), for a married pair.) Since Bobby is married to Jerry, Kim's statement 6 is true.

Since Bobby is married to Jerry, we see that Ronnie's statement 12 is false. We have now identified 4 persons who lie, namely, Pat, Willie, Freddie, and Ronnie. Since two liars cannot be married, we conclude that Ronnie and Willie are not married. Now, from the table it is clear that Freddie is married to either Kim or Lou. Suppose that Freddie were married to Kim: then Terry's statement 13 is false which implies that Terry is the fifth and last liar; but statements 2 and 4 are contradictory, which implies that either Ellie or Jerry would be yet another liar, which is impossible. The only possible conclusion from this is that Freddie is married to Lou. By elimination we have Kim married to Pat. Also by elimination we have Ellie married to Willie and Ronnie married to Terry.

Since Ellie is married to Willie, the doctor, who lies, Ellie must tell the truth; this implies that Pat is the rich man. Since Jerry tells the truth, Freddie is the beggar man. Since Ronnie lies, Jerry must be a man. But Jerry tells the truth; so Jerry is the poor man. The fifth man must belong to the couple Ronnie-Terry; he must be the thief and thus is a liar. This forces the conclusion that Ronnie is the thief.

(125) *Problem.*   Three persons are shown 3 red hats and 2 black hats. They are then seated in chairs placed in single file and blindfolded. A hat is placed on each person's head, the remaining hats hidden, and the blindfold removed. One at a time, they are asked if each can guess the colour of the hat on his own head.

The person who sits in the third chair is asked first, and he confesses that he does not know the colour of his hat, even though he can see the hats on the heads of his two companions. The second person, who can see only the hat on the person in front of him, also admits he cannot guess his colour. The first person, who can see no hats at all, says he knows the colour of his hat and he is correct.

What colour hat is he wearing, and how did he know?

Solution.   If the first person saw 2 black hats he would know his own was red since there were only 2 black hats available. Since he says he does not know the colour of his own hat, then he must see either 2 red hats or a red hat and a black hat. Person 2 logically concludes from the first person's negative response exactly what we have just concluded. Thus if person 2 saw a black hat he would know that his own hat must be red. Since he does not know the colour of his own hat, he must see a red hat on person 3. Person 3 reasons in this manner to arrive at the same conclusion.

Note that no matter the response of person 2, person 3 can still deduce the colour of his hat, and if person 1 is able to deduce the colour of his hat, so can the others deduce the colour of their hats.

(126) *Problem.*   Sixteen passengers on a liner discover that they are an exceptionally representative body. Four are English, four are Scots, four are Irish, and four are Welsh. There are also four each of different ages: 35, 45, 55, and 65; and no two of the same age are of the same nationality. By profession also, four are lawyers, four soldiers, four doctors, and four clergy, and no two of the same profession are of the same age or of the same nationality.

It appears also that four are single, four married, four widowed, and four divorced, and that no two of the same marital status are

of the same profession, or the same age, or of the same nationality.
Finally, four are conservatives, four liberals, four socialists, and
four fascists, and no two of the same political sympathies are of
the same marital status, or the same profession, or the same age,
or the same nationality.

Three of the fascists are known to be a single English lawyer of
65, a married Scot soldier of 55, and a widowed Irish doctor of 45.
It is then easy to specify the remaining fascist. It is further given
that the Irish socialist is 35, the conservative Scot is 45, and the
Englishman of 55 is a clergyman. What do you know of the Welsh
lawyer?

Solution.   It is easy to determine that the remaining fascist must
be a divorced clergy of 35. Let us now create a 4 × 4 grid to contain
the known information.

	Fascist	Conservative	Liberal	Socialist
English	lawyer single 65			
Scot	soldier married 55	45		
Irish	doctor widowed 45			35
Welsh	clergy divorced 35			

The rest of this grid must be filled in with each of the four profes-
sions, each of the four marital statuses, and each of the four ages
in each row and column.

There are two more 35 year olds to place in the grid. They must be
English and Scot, and also must be conservative and liberal. Since
the conservative Scot is 45, we deduce that the two remaining 35
year olds are an English conservative and a liberal Scot. This
further implies that the socialist Scot is 65. This socialist Scot

must be either a doctor or clergy and must be either widowed or divorced, from which we conclude that she must be a divorced doctor or a widowed clergy. But we are told that the widowed clergy is English and 55. Therefore, the socialist Scot is a divorced doctor (of 65). It now follows that the conservative Scot is a single clergy (of 45) and the liberal Scot is a widowed lawyer (of 35).

We now must have the widowed English clergy of 55 as a socialist, which implies that the English liberal is 45 as is the Welsh socialist. The next conclusion is that Irish socialist must be a single soldier (of 35), which makes the Welsh socialist a married lawyer (of 45). Thus the Welsh lawyer is a married socialist of 45. This completes the solution.

However, more conclusions follow rapidly. The English liberal is a divorced soldier (of 45) and the English conservative is a married doctor (of 35), and the rest of the grid must be completed as shown below:

	Fascist	Conservative	Liberal	Socialist
English	lawyer single 65	doctor married 35	soldier divorced 45	clergy widowed 55
Scot	soldier married 55	clergy single 45	lawyer widowed 35	doctor divorced 65
Irish	doctor widowed 45	lawyer divorced 55	clergy married 65	soldier single 35
Welsh	clergy divorced 35	soldier widowed 65	doctor single 55	lawyer married 45

(127) *Problem.*   Consider three circles in the plane that intersect to form seven bounded regions. In each region there is a token that is white on one side and black on the other. At any stage the following two operations are permissible: (a) we can invert (flip over) all four tokens inside one of the three circles, or (b) we can invert those tokens showing black inside one of the three circles

so that afterwards all tokens in that circle show white. From the starting configuration in which all tokens show white, can we reach the configuration in which all tokens show white except that the central region common to all three discs shows black?

Solution. Call a configuration "all-odds" if each of the three circles contains an odd number of black tokens. In particular, the desired ending configuration (in which all regions but the central one show black) is all-odds. Since an operation of type (a) flips either two or four tokens inside each of the three circles, it does not change the parity of the number of black tokens in any circle. On the other hand, an operation of type (b) results in an even number of black tokens in at least one circle and so the use of (b) at any stage precludes the possibility of ending up with an all-odds configuration. Hence, an all-odds configuration can be the end result only when we have made merely operations of type (a) upon another all-odds configuration. Since the given initial configuration (all white) is not all-odds, the desired configuration cannot be reached from it.

(128) *Problem.* The ruler of the ancient Greek island of Chios was explaining to two goldsmiths that he had come into a little money, and wanted the new monumental pillar changed.

It was cylindrical, and he wanted it covered with solid gold, but the gold was to form four equal hemicylindrical projecting mouldings, the plan being as shown below.

Where these four surfaces met they were to form right angles with one another, and exactly at the surface of the existing pillar.

One of the goldsmiths was rather deaf, and thought what was desired was a new solid square pillar, the four corners of which were to fit exactly in the cylindrical surface of the one it was to replace. The goldsmiths were then asked to estimate the amount of gold needed.

To what extent did their estimates differ? We are to assume that both calculations were correct on the basis of what each thought were the requirements.

Solution. They were the same. The proof is as follows: Let us consider only one of the four hemicylindrical covers, *a*, and one quarter of the square replacement pillar, *ABC*, and demonstrate their equality.

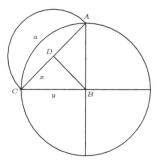

D is the mid-point of the side *AC*, and we make $x = 1$. Then $y = \sqrt{2}$.

$$\text{Area of } a = \frac{\pi}{2} - \left(\frac{\pi\sqrt{2}^2}{4} - \frac{\sqrt{2}^2}{2} \right)$$

$$= \frac{\pi}{2} - \frac{2\pi}{4} + \frac{2}{2} = 1.$$

$$\text{Area of } ABC = \frac{\sqrt{2}^2}{2} = 1.$$

(129) *Problem.* A digital clock forms palindromic numbers (numbers that read the same both forward and backward) 114 times each day. What is the least amount of time between two palindromic numbers? What is the most?

Solution. First notice that there are many times when two palindromic times are exactly 10 minutes apart (e.g. 1:11 to 1:21). Indeed this occurs regularly when the time is expressed using 3 digits. The smallest possible amount of time might be as small as 1 minute. This would imply that the last digit and the leading digit would both have to change, which could only happen if the last digit changed from 9 to 0 with the subsequent carry moving left to the leading digit which would also change from 9 to 0 (but a leading digit of 0 is never written). This argument shows that we must have at least 2 minutes between two successive palindromic times.

However, we can use the same argument to show that when we look for two palindromic times differing by two minutes we must again have the last digit advance from 9 to 1 with the carry left to the leading digit; interestingly enough this does give rise to a new palindromic time: that is, between 9:59 and 10:01 there are two minutes.

When we have 4-digit times we no longer get palindromes every 10 minutes, but only every 1 hour and 10 minutes (that is, 70 minutes). There are only three 4-digit palindromic times, namely 10:01, 11:11, and 12:21. Thus, 70 minutes is the longest separation and it occurs twice: from 10:01 to 11:11 and from 11:11 to 12:21.

(130) *Problem.*   A bag contains 17 marbles. They are of four different colours; there are at least two of each colour; and there are differing numbers of marbles for each of the four colours.

There are more green marbles than there are of any other colour. If I draw from the bag enough marbles—but only just enough—to ensure that I have at least two marbles in each of two different colours, I must draw 11 marbles.

How many must I draw to ensure that at least one of the marbles drawn is green?

Solution.   From the information in the first paragraph of the problem statement, we can conclude that there are only three possible colour distributions:

(a)   2 3 4 8
(b)   2 3 5 7
(c)   2 4 5 6

The last number in each row must represent the number of green marbles. To ensure that at least two marbles in each of two different colours are drawn requires a different number of draws in the three cases. In case (a) 12 are needed; in case (b) 11 are needed; in case (c) 10 are needed. Hence the actual number of green marbles is 7. Thus, the number of draws required to be sure of drawing at least one green marble is again 11.

(131) *Problem.*   The table on the next page shows some of the results of last summer's Frostbite Falls Fishing Festival, showing how many contestants caught n fish for various values of n:

n	0	1	2	3	$\cdots$	13	14	15
number of contestants who caught n fish	9	5	7	23	$\cdots$	5	2	1

In the newspaper story covering the event, it was reported that

    (a) the winner caught 15 fish;

    (b) those who caught 3 or more fish averaged 6 fish each;

    (c) those who caught 12 or fewer fish averaged 5 fish each.

What was the total number of fish caught during the festival?

Solution. Let C be the number of contestants who caught between 4 and 12 fish, inclusive, and let F be the number of fish caught by those contestants. Since those who caught 3 or more fish averaged 6 fish each, we get

$$6(23 + C + 5 + 2 + 1) = 69 + F + 65 + 28 + 15$$
$$\text{or} \quad 6C + 186 = F + 177,$$
$$\text{that is,} \quad 6C = F - 9.$$

Since those who caught 12 or fewer fish averaged 5 fish each, we see that

$$5(9 + 5 + 7 + 23 + C) = 0 + 5 + 14 + 69 + F$$
$$\text{or} \quad 5C + 220 = F + 88,$$
$$\text{that is,} \quad 5C = F - 132.$$

From these two equations we immediately have $C = 123$, which implies that $F = 747$. Then, the total number of fish caught is

$$0 + 5 + 14 + 69 + F + 65 + 28 + 15 = F + 196 = 943.$$

(132) *Problem.* Starship Venture is under attack from a Zokbar fleet, and its Terrorizer is destroyed. While it can hold out, it needs a replacement to drive off the Zokbars. Starbase has spare Terrorizers, which can be taken apart into any number of components, and enough scout ships to provide transport. However, the Zokbars have n Space Octopi, each of which can capture one scout ship at a time. Starship Venture must have at least one copy of each component to reassemble a Terrorizer, but it is essential that the Zokbars should not be able to do the same. Into how many components must each Terrorizer be taken apart (assuming all are taken apart in an identical manner), and how many scout ships are needed to transport them? Consider two situations.

(a) Assuming that the number of components per Terrorizer is as small as possible, minimize the number of scout ships.

(b) Assuming instead that the number of scout ships is as small as possible, minimize the number of components per Terrorizer.

Solution. Note first that the minimum number of spare Terrorizers Starbase must have is $n + 1$. If there are only n of them, there are only n copies of each component. If the Zokbars capture n scout ships carrying a common component, Starship Venture will not get a complete Terrorizer. On the other hand, if there are $n + 1$ spare Terrorizers, Starship Venture will get at least one copy of each component, as long as no two copies of any component are ever carried by any one scout ship. Hence there is no advantage in disassembling more than $n + 1$ spare Terrorizers. We will also assume that no two copies of any component are ever carried by any one scout ship.

(a) The minimum number of components into which each Terrorizer must be taken apart is $n + 1$. Otherwise, the Zokbars may get a complete Terrorizer by capturing, for each component, one scout ship that is carrying it. If each Terrorizer is taken apart into $n + 1$ components, Starbase must have $(n + 1)^2$ scout ships. Each will carry one component of one Terrorizer. If there are fewer scout ships, then some scout ship must carry at least two (different) components. The Zokbars can capture this scout ship and, for each of the other components, one scout ship carrying it.

(b) Starbase must have at least $2n + 1$ scout ships. Otherwise, the Zokbars will end up capturing a number of scout ships equal to that which get through to Starship Venture. If one side can get a complete Terrorizer, so can the other. If Starbase has exactly $2n+1$ scout ships for this mission, each Terrorizer must be taken apart into $\binom{2n+1}{n}$ components, one for each subset of n of the $2n+1$ scout ships. Each component will be carried by the $n+1$ scout ships not in the subset associated with it. Whichever n scout ships the Zokbars capture, exactly one component will be missing. On the other hand, if the number of components is smaller, two distinct (not necessarily disjoint) subsets of n scout ships (that is, at least $n + 1$ scout ships altogether) will then be missing the same component. This is impossible since there are $n + 1$ copies of each component.

(133) *Problem.* Both the Allens and the Smiths have two young sons under the age of eleven (rounded to the nearest year). The names of the boys, whose ages rounded off to the nearest year are all different, are Arthur, Bert, Carl, and David. Taking the ages of the boys only to the nearest year, the following statements are true.

(a) Arthur is three years younger than his brother.
(b) Bert is the oldest.
(c) Carl is half as old as one of the Allen boys.
(d) David is five years older than the younger Smith boy.
(e) The total ages of the boys in each family differ by the same amount today as they did five years ago.

How old is each boy, and what is each boy's family name?

Solution. Let A, B, C, and D be the ages of Arthur, Bert, Carl, and David, respectively. It is clear that $B > D \geq 5$, and that $A, B, C, D \leq 10$, which further implies that $C \leq 5$. We also have either $D = C + 5$ or $D = A + 5$.

Suppose first that $D = A + 5$. Then Arthur is the younger Smith, and the first statement says that Arthur's brother is 3 years older; this clearly eliminates both Bert and David.

Therefore, we would have Carl Smith and Arthur Smith (both 5 or less) and Bert Allen and David Allen (both at least 5). Five years ago only Bert and David would have had a positive age; their age total would have been 10 less than today, which would imply that the sum of the ages of Arthur and Carl would have to be 10, which is impossible since both are different and at most 5.

Thus, $D = C + 5$, which implies that Carl's family name is Smith, and that he is the younger of the two Smith boys. Furthermore, we see that Arthur must be the younger Allen.

Let us now suppose that Bert is an Allen and David is a Smith. Then $B = A + 3$, and either $2C = B$ or $2C = A$. If $2C = B$, then $2C > D = C + 5$ which implies that $C \geq 6$, which is impossible. Thus, $2C = A$. In this case we get $10 \geq B = 2C + 3$ and $2C + 3 = B > D = C + 5$, from which we can deduce that $C = 3$, which yields $C = 3$, $D = 8$, $A = 6$, and $B = 9$ which contradicts the last sentence in the problem statement true. Therefore, we conclude that Bert is a Smith and David is an Allen.

Thus, we have $B > D = C + 5 = A + 3$ and either $2C = A$ or $2C = D$. If $2C = D = C + 5$, then $C = 5$, $D = 10$, and $B > 10$, which is impossible.

Therefore, $2C = A$, which from $C + 5 = A + 3$ implies that $C = 2$, $A = 4$, and $D = 7$. The current difference in the sum of the ages of the boys from the two families is $(4 + 7) - (2 + B) = 9 - B$. Five years ago Carl and Arthur were not born, so that the difference was $7 - B$. Clearly these values cannot be equal. However, it states that the difference must be the same, so that these two differences could be negatives of each other and still satisfy the condition. In this case (that is, $9 - B = B - 7$), so that we get $B = 8$.

Thus, Arthur Allen is 4, Bert Smith is 8, Carl Smith is 2, and David Allen is 7.

(134) *Problem.*   On the night of October 22, the Vancouver home of I. M. A. Richman was robbed and $80,000 worth of jewelry was taken.

The following five suspects were questioned and, from their statements, the robber was found and arrested. Knowing that each man made *only one* false statement and one of these five is the burglar, can you also pick him out? Use only logic, reason, and common sense.

   (a) *Andy*: I was in Kamloops the night of the robbery. Butch is the guilty man. Charlie lied when he said the Dope did it. I am innocent.

   (b) *Butch*: Andy lied when he said I did it. I am innocent. Charlie is not telling the truth about being in Kamloops. I never robbed anyone in my life.

   (c) *Charlie*: The Dope did it because he told me so. I am not the guilty man. I was in Kamloops with Andy the night of the robbery. Butch is innocent.

   (d) *The Dope*: Butch lied when he said Charlie is not telling the truth. I am innocent. Charlie is the man who did it. Ed was in Richmond on October 22.

   (e) *Ed*: I am absolutely innocent. Andy is innocent too. I was in Richmond on October 22. Charlie was in Kamloops with Andy the night of the robbery.

Solution.   If Butch was guilty, his statements "I am innocent" and "I never robbed anyone in my life" would both be false, which is impossible. Therefore, Butch is innocent. Thus, Andy's only false statement is "Butch is the guilty man". Therefore, Andy was in

Kamloops the night of the robbery, and is thus, innocent. Andy's other statement must also be true, which implies that the Dope is innocent. This means that Charlie's first statement is his only false statement, and thus, Charlie is also innocent. This leaves Ed as the guilty man, and one can check that all the conditions are met in this case.

(135) *Problem.*   Prince Ivan made up his mind to fight the three-headed, three-tailed dragon. So he obtained a magic sword that could, in one stroke, chop off either one head, two heads, one tail, or two tails. A witch revealed that dragon's secret to him: if one head is chopped off, a new head grows; in place of one tail, two new tails grow; in place of two tails, one new head grows; and if two heads are chopped off, nothing grows. What is the smallest number of strokes Prince Ivan needs to chop off all the dragon's heads and tails?

Solution.   The answer is nine. After Prince Ivan makes h_i $(i = 1, 2)$ strokes that chop off i of the dragon's heads and t_j $(j = 1, 2)$ strokes that chop off j tails, the number of heads and tails remaining will be $3 - 2h_2 + t_2$ and $3 + t_1 - 2t_2$, respectively. We wish to find the solution where both of these are 0, which also has minimum value for $h_1 + h_2 + t_1 + t_2$. Clearly, we must set $h_1 = 0$. Thus, we must have

$$2h_2 - t_2 = 3,$$
$$2t_2 - t_1 = 3.$$

From the first equation we have t_2 odd; from the second equation we see that $t_2 \geq \frac{3}{2}$, so $t_2 \geq 3$. Then $h_2 = (3 + t_2)/2 \geq 3$ and $t_1 = 2t_2 - 3 \geq 3$, so that the total number of strokes is at least 9. The numbers $h_2 = t_1 = t_2 = 3$ satisfy our equations; we must now show that it is possible to actually do the job with 3 strokes of each of the 3 types (so that, for example, Ivan would not have to chop off two tails when there is only one tail left). A possible sequence is to chop off one tail (so that two new heads grow), then two tails (so that one new head grows), then two heads, and then repeat this series of strokes twice; after each of the three series the dragon loses one tail and one head.

(136) *Problem.*   The results were to be announced in the annual Ruritanian song festival, but there was apparently some delay. Gradually the word was passed round that one of the four finalists, instead of

giving three marks to the other finalist he rated highest, and two marks for the next best finalist and one mark for the third best (naturally no finalist was asked to rate his own song), had reversed the marks given, hoping to favour his own chances. He had given one mark to his best choice, two marks to his middle choice and three marks to the finalist he actually thought worst. The commotion was of course tremendous and it only increased when it was revealed that two of the other finalists had taken exactly the same dishonest step in the hope, so they thought, of improving their chances. Before these revelations were made all four finalists had been tied on six points. When the judges eventually reversed the marking orders of the three dishonest finalists, in what place did the honest singer find himself?

Solution. The original scores of 6 points to each finalist must have come from a first, second, and a third placing each. When the three dishonest singers reversed their order or marking, the 3 points and 1 point scored by the honest singer was reversed, but his total remained at 6. The other three finalists, however, when the marks were reversed, gained 2, lost 2, or stayed the same, depending on whether they scored 1, 2 or 3 from the honest singer. Thus, the final scores were 8, 6, 6, and 4, and the honest man was tied for second.

(137) *Problem.*   At 9:10 pm Charley left Belamy's Tavern to go to the Galaxy Pub. At exactly the same time, Harry left the Galaxy to go to Belamy's. Each walked a direct route at his own constant rate. They passed each other at 9:24 pm, but since they were not acquainted, they did not speak. At their new destinations, each spent exactly 25 minutes quenching his thirst, then headed straight back to his original bar at the same pace as before. At what time did they pass each other again?

Solution.   Call the distance between the bars d. Let us first ignore the 25 minutes not travelling. Although their individual speeds may have been different, Charley and Harry together covered the distance d between the two bars in 14 minutes. When they met a second time, they had together traveled an additional distance of $2d$, so it took an additional 28 minutes. Since they did spend 25 minutes not travelling, the second crossing of the paths took place $28 + 25 = 53$ minutes after the first crossing, at 10:17 pm.

(138) *Problem.* We have Xed out all but a few of the digits in the long division problem below. Using mathematical logic, can you replace the Xes with numbers, one digit per X, so that the completed division is correct? Find the complete solution, not just the answer.

$$
\begin{array}{r}
X X X X X X \\
X X \overline{)\, X X 1 X X X 9 X} \\
\underline{X X X} \\
X X X \\
\underline{X\,6} \\
6 X X \\
\underline{X X X} \\
X X 9 \\
\underline{X 4 X} \\
X X X \\
\underline{X X X} \\
0
\end{array}
$$

Solution. Let us number the lines below the dividend from 1 to 9. Now it is clear that the second digit of the quotient must be 0. Next notice that the leading digit of the divisor must be at least 6 since it does not divide into $6X$, but does divide into $6XX$. This further implies that the third digit of the quotient must be 1, since twice the divisor must have at least 3 digits, which would contradict line 3. Thus, the units digit of the divisor must be 6. Therefore, the divisor is one of 66, 76, 86, or 96. Of these the only one which has a multiple (by a single digit) with a 4 in the tens position (see line 7) is 86.

Thus, the divisor is 86, the tens digits of the quotient is 4 and line 7 is 344. The leading digit of line 2 must be 1, whence, the units digit of line 2 is 0, implying the leading digit of the quotient is 5, further implying that line 2 is 430 and the dividend begins with 431. Since the leading digit of line 6 is either 3 or 4, we see that the only multiple of 86 which will work for line 5 is 602. Thus, line 5 is 602 and the hundreds digit of the quotient is 7. Now the tens digit of lines 8 and 9 is 5, and the only multiple of 86 by a single digit which has a 5 in the tens position is 258.

Therefore, lines 8 and 9 are both 258 and the quotient is now completely determined to be 501743. This also yields the units digit of the dividend as 8. Lines 8 and 7 tell us that line 6 is 369.

This and line 5 give us that line 4 is equal to 638. Carrying this one more step gives line 2 as 149, yielding the dividend 43149898. Therefore, the final answer is:

$$
\begin{array}{r}
5\,0\,1\,7\,4\,3 \\
8\,6\,)\overline{4\,3\,1\,4\,9\,8\,9\,8} \\
\underline{4\,3\,0} \\
1\,4\,9 \\
\underline{8\,6} \\
6\,3\,8 \\
\underline{6\,0\,2} \\
3\,6\,9 \\
\underline{3\,4\,4} \\
2\,5\,8 \\
\underline{2\,5\,8} \\
0
\end{array}
$$

(139) *Problem.*   A game of Jai Alai has eight players and starts with players P_1 and P_2 on the court and the other players P_3, P_4, P_5, P_6, P_7, and P_8 waiting in a queue. After each point is played, the loser goes to the end of the queue; the winner adds 1 point to his score and stays on the court; and the next player at the head of the queue comes on to contest the next point. Play continues until someone has scored 7 points. At that moment, we observe that a total of 37 points have been scored by all eight players. Determine who has won.

Solution.   Suppose that P_k is the winner. Then he must win 7 games and the sum of the other players' scores must be 30. It takes 6 turns for a player to go from the back of the queue to the court and therefore each time P_k loses, the sum of the other scores increase by 7 before P_k returns to the court. Since P_k is the winner, he must win the final game, and

$$30 = 7(\text{number of times } P_k \text{ loses})$$
$$+ (\text{number of games before } P_k \text{ gets onto the court}$$
$$\text{for the first time}).$$

Since $30 = 7 \times 4 + 2$, there must be 2 games played before the (eventual) winner takes the court for the first time. Hence, P_4 is the winner.

(140) *Problem.* What is the most far-reaching conclusion that one can reach from the following set of premises due to Lewis Carroll?

1. All policemen on this beat sup with our cook.
2. No man with long hair can fail to be a poet.
3. Amos Judd has never been in prison.
4. Our cook's 'cousins' all love cold mutton.
5. None but a policeman on this beat are poets.
6. None but her 'cousins' ever sup with our cook.
7. Men with short hair have all been in prison.

Solution. There are a number of sets of people that are described in these premises. Let us designate the sets by:

$$A = \{\text{policemen on this beat}\}$$
$$B = \{\text{those who sup with our cook}\}$$
$$C = \{\text{persons with long hair}\}$$
$$D = \{\text{poets}\}$$
$$E = \{\text{Amos Judd}\}$$
$$F = \{\text{those who have been in prison}\}$$
$$G = \{\text{our cook's 'cousins'}\}$$
$$H = \{\text{those who love cold mutton}\}$$

We will use the notation X^c to represent the complement of a set X. Let us translate the English statements in the problem into set language:

$$1 : A \subseteq B$$
$$2 : C \subseteq D$$
$$3 : E \subseteq F^c$$
$$4 : G \subseteq H$$
$$5 : D \subseteq A$$
$$6 : B \subseteq G$$
$$7 : C^c \subseteq F$$

The last relation is equivalent to $F^c \subseteq C$. This allows us to string the relations together to get:

$$E \subseteq F^c \subseteq C \subseteq D \subseteq A \subseteq B \subseteq G \subseteq H,$$

from which we conclude $E \subseteq H$, that is, Amos Judd loves cold mutton.

(141) *Problem.*   My only timepiece is a wall clock. One day I forgot to wind it and it stopped. I went to visit a friend whose watch is always correct, stayed awhile, and returned home. There I made a simple calculation and set the clock right.

How did I do this when I had no watch on me to tell how long it took me to return from my friend's house? (I also have no access to telephone or any electronic media such as radio or television.)

Solution.   Before I left I wound the wall clock. When I returned, the change in time it showed equalled the time it took to go to my friend's house and return, plus the time I spent there. But I knew the latter because I looked at my friend's watch when I arrived and when I left.

Subtracting the time of the visit from the time I was absent from my house, and dividing by 2, I obtained the time it took me to return home. I added this time to what my friend's watch showed when I left, and set that sum on my wall clock.

(142) *Problem.*   On the blackboard, there are written the numbers 48, 24, 16, ..., $\frac{48}{97}$; that is, rational numbers $\frac{48}{k}$ with $k = 1, 2, \ldots, 97$. In each step two arbitrarily chosen numbers, a and b, on the blackboard are erased and the number $2ab - a - b + 1$ is written on the blackboard in their place. After 96 steps there is only one number on the blackboard. Determine the set of all possible outcomes of the procedure.

Solution.   The only possible outcome is $\frac{1}{2}$.

Let E_n be the set of all numbers remaining on the blackboard after the nth step for $n = 0, 1, \ldots, 96$. Note that $\frac{1}{2} = \frac{48}{96} \in E_0$, where E_0 is the initial set of numbers.

Now let k be any fixed integer, $0 \leq k \leq 95$. Suppose that $\frac{1}{2} \in E_k$. In the $k + 1^{\text{st}}$ step, we choose a and b. If neither a nor b is $\frac{1}{2}$, then $\frac{1}{2}$ remains in E_{k+1}. On the other hand, if $a = \frac{1}{2}$, then $2ab - a - b + 1 = \frac{1}{2} \in E_{k+1}$. Similarly, if $b = \frac{1}{2}$, we get $\frac{1}{2} \in E_{k+1}$. Therefore, no matter the choice of a and b, we have $\frac{1}{2} \in E_{k+1}$, and, by induction, we conclude that $\frac{1}{2} \in E_n$ for all n, $0 \leq n \leq 96$. Since E_{96} contains only one number, that number must be $\frac{1}{2}$, as claimed.

(143) *Problem.*   A set of (double-6) dominoes consists of 28 rectangular tiles. Each tile has two squares, each with a 0, 1, 2, 3, 4, 5, or 6. Every combination is represented. The value of the tile is the sum of the values of the two squares. When both squares have the same value, the tile is called a doublet.

The basic rule in playing dominoes is that, in adding to a chain, you have to match the value of one square of your tile to the value of a square at one end of the chain. A game is won by the team which first has a player to play his/her last tile. If a game is deadlocked (that is, no one has a playable tile), victory goes to the team with the smallest total on their remaining tiles.

Suppose that Anne and Chuck are playing Bill and Diane. Each player has 6 tiles, and 4 tiles remain face down, and are not used. Anne has: 0-4, 1-3, 1-4, 1-5, 2-3, 2-4.

Her partner, Chuck, has 5 doublets. Diane has 2 doublets; her tiles total 59 points.

Anne plays 2-4; Bill passes (a player may only pass if he has no tile that can be played); Chuck plays a tile; Diane passes; Anne plays a second tile; Bill passes again; Chuck plays a tile, and the game is deadlocked; that is, all four players must pass. Anne and Chuck together have 35 points left on their tiles, while Bill and Diane have 91 points left. The four tiles played had a value of 22 points.

Which 4 tiles were left face down, and which four tiles were played? *Solution.* Let us make some preliminary observations. First we note that each number from 0 to 6 appears a total of 8 times among the 28 tiles (once with each of the other 6 numbers, and once in a doublet). Thus, the total value of all the tiles is

$$8(0 + 1 + 2 + 3 + 4 + 5 + 6) = 8(21) = 168.$$

Bill and Diane have a total of 91 (both before and after play, since neither played a tile); this leaves a total of 77 points. At the end of play Anne and Chuck have 35 points remaining, and together they played a total of 22 points. Thus, the four face down (unused tiles) have a total of

$$77 - 35 - 22 = 20$$

points. Since the tiles played account for 22 points, and Anne initially played a tile worth 6 points, the remaining 3 tiles played have a total value of 16 points.

Since Bill could not play, we conclude that Bill has no tile with either a 2 or a 4 on it. Since there are only 7 doublets, and Chuck and Diane have 7 doublets between them, we also know that Bill has no doublets.

Thus, Bill's six tiles must be chosen from among the eight tiles:

0-1, 0-3, 0-5, 0-6, 1-6, 3-5, 3-6, 5-6.

If Chuck (on his first play) played either of the doublets 2-2 or 4-4,
the ends of the chain would still be a 2 and a 4; then since Diane
passed, we could conclude that Diane had neither a 2 or a 4 on any
of her tiles.

Thus, her 4 non-doublet tiles would have to be selected from among
the 8 tiles listed above, which is impossible since Bill needs 6 of
them himself. Therefore, we know that Chuck did not play a doublet on his first play.

Since Chuck made a play, the tile he first played must have either
a 2 or a 4 on it, and is not a doublet. We will consider possibilities
that flow from each such first play of Chuck, recalling that the
total of the played tiles after Anne's initial play is 16, and that
after Chuck's first play, he has only doubles left, which means the
total value of his remaining tiles is even from this point on.

Chuck plays	Chain ends	Sum of tiles still to play	Anne's next play	Chuck's next play	
$0-2$	$0, 4$	14	$0-4$	$5-5$	(which won't play)
			$0-4$	$1-4$	no play (odd)
$1-2$	$1, 4$	13	$0-4$		no play (odd)
			$1-3$		no play (odd)
			$1-5$		no play (odd)
			$1-4$	$4-4$	Anne can still play $0-4$
$2-5$	$4, 5$	9	$0-4$		no play (odd)
			$1-5$		no play (odd)
			$1-4$	$2-2$	(which won't play)
$2-6$	$4, 6$	8	$0-4$	$2-2$	(which won't play)
$3-4$	$2, 3$	9	$1-3$		no play (odd)
			$2-3$	$2-2$	
$4-5$	$2, 5$	7	$1-5$		no play (odd)
			$2-3$	$1-1$	(which won't play)
$4-6$	$2, 6$	6	$2-3$		no play (odd)

As seen in the table above the only possibility is for Chuck to
play 3-4, Anne to play 2-3, and finally Chuck to play 2-2. This
answers the second part of the question asked, namely which tiles
were played.

Anne is now left with the tiles 0-4, 1-3, 1-4, and 1-5, which have
a total value of 19 points, which means that Chuck has a total
of $35 - 19 = 16$ points, and all of Chuck's remaining 4 tiles are
doubles, excluding 2-2, which he has already played.

The four smallest remaining doubles are 0-0, 1-1, 3-3, and 4-4, which sum to 16; thus, Chuck must have these tiles. The tiles 0-2, 1-2, 2-5, and 2-6 did not belong to either Anne or Chuck, since we now know precisely which tiles they had. Neither do they belong to Bill or Diane, since they could be played.

Therefore, these are the 4 tiles which were face down and not played.

(144) *Problem.*   A magician has one hundred cards numbered 1 to 100. She puts them into three boxes, a red one, a white one, and a blue one, so that each box contains at least one card. A member of the audience selects two of the three boxes, chooses one card from each and announces the sum of the numbers on the chosen cards. Given this sum, the magician identifies the box from which no card has been chosen. How many ways are there to put all the cards into boxes so that this trick always works? (Two ways are considered different if at least one card is put into a different box.)

Solution.   We first claim that 1 and 2 are in different coloured boxes. If not, say $1, 2, \ldots, i-1$ are in red, i is in white, and let j be the smallest number in the blue box. We have $i \geq 3$ and $j - 1 \geq i$. But in view of

$$i + (j - 1) = (i - 1) + j,$$

we conclude that $j - 1$ is in the same coloured box as i, namely white, which leads to the fact that the sum

$$2 + (j - 1) = 1 + j$$

does NOT allow the magician to decide on the unpicked box.

Now let 1 be in the red box, 2 be from the white box, and j be the smallest number in the blue box. We consider the following cases:

(a) $j = 3$. Since $1 + 4 = 2 + 3$, we see that 4 must be from the red box. Similarly, 5 is from the white box, 6 is from the blue box, and so on cyclically.

(b) $j = 100$. Since $2 + 99 = 1 + 100$, we see that 99 is from the white box. If $t > 1$ is from the red box, since

$$t + 99 = (t - 1) + 100,$$

we conclude that $t - 1$ is from the blue box, but 100 is the smallest number from the blue box. Therefore, $2, 3, \ldots, 99$ are all white.

(c) $3 < j < 100$. Since
$$2 + j = 1 + (j + 1),$$
we conclude that $j + 1$ is from the red box. Since
$$3 + j = 2 + (j + 1),$$
we also conclude that 3 is from the blue box, but j is the smallest number from the blue box, a contradiction.

Therefore, there are three choices of colours for number 1, two choices of colour for number 2, and two choices for number 3. Once these choices are made, the colours for the remaining numbers are determined. Thus, the answer to this problem is $3 \times 2 \times 2 = 12$.

(145) *Problem.*   Five pirates find a cache of five gold coins. They decide that the shortest pirate will become bursar and distribute the coins — if half or more of the pirates (including the bursar) agree to the distribution, it will be accepted; otherwise, the bursar will walk the plank and the next shortest pirate will become bursar. This process will continue until a distribution of coins is agreed upon. If each pirate always acts so as to stay aboard if possible and maximize his wealth, and would rather see another walk the plank than not (all else being equal), then how many coins will the shortest pirate keep for himself?

Solution.   Call the pirates p_1, p_2, p_3, p_4, p_5, with p_1 the shortest up to p_5 the tallest. Consider what would happen if only p_4 and p_5 remained. Whatever division strategy p_4 suggested would hold, since p_4's vote alone would be half the total vote. Thus, p_4 would simply allot himself all the gold, and p_5 would get nothing. Next consider the situation where three pirates remained. Whatever distribution p_3 chose, p_5 would have to agree, as long as he got one or more coins. Otherwise, he would go to the 2-pirate situation and he would get nothing! Hence p_3 would simply take 4 coins himself and allot 1 coin to p_5, getting a majority vote from himself and p_5. Similarly, with four pirates remaining and p_2 as bursar, p_2 would take 4 coins for himself and allot 1 coin to p_4. Again, since p_4 would have otherwise received nothing with p_3 as bursar, he would have to support the plan, which would then have a majority vote. In the 5-pirate situation, the bursar, p_1, would allot himself 3 coins and give one each to p_3 and p_5. Since each of p_3 and p_5 would receive nothing with p_2 as bursar, they would have to vote in favour of this plan. Thus, the shortest pirate gets 3 coins.

(146) *Problem.* In a certain multiple-choice test, one of the questions was illegible, but the choice of answers, given below, was clearly printed. What is the right answer?

(a) All of the below.

(b) None of the below.

(c) All of the above.

(d) One of the above.

(e) None of the above.

(f) None of the above.

Solution. Note first that if (f) is true, then so is (e), which implies that (f) is false, which is a contradiction. Thus, (f) is false. Since (f) is false, so is (a), which further implies that (c) must also be false. If (b) is true, then so is (d), which contradicts (b) being true. Thus, (b) is false. Since we now have (a), (b), and (c) false, we conclude that (d) is false, which means that (e) is true. Therefore, (e) is the right answer.

(147) *Problem.* Twenty-one girls and twenty-one boys took part in a mathematical competition. It turned out that

(a) each contestant solved at most 6 problems, and

(b) for each pair of a girl and a boy, there was at least one problem that was solved by both the girl and the boy.

Prove that there is a problem that was solved by at least three girls and at least three boys.

Solution. Assign each problem a unique letter, and also number the boys $1, 2, \ldots, 21$ and number the girls $1, 2, \ldots, 21$. Construct a 21×21 matrix of letters as follows: in the ith row and j^{th} column, write the letter of all problems that both the ith girl and the jth boy solved—at least one such problem exists for each ordered pair (i, j) by condition (b). If we consider the ith row, each letter in that row corresponds to a problem that the ith girl solved. Since each girl solved at most six problems, each row contains at most 6 distinct letters. Similarly, each column contains at most 6 distinct letters.

We have the following key observation: *In each row, consider the letters which appear at least three times. At least 11 squares in the row contain one of these letters.* (A similar observation can be made about the columns.) Indeed, there are at most 6 different letters, and they cannot all appear at most twice, since that would account for letters in at most 12 of entries in that row or column, and we know there is at least one letter in each of the 21 entries. So at most 5 different letters appear at most twice, giving a total of at most 10 squares containing letters appearing at most twice. Then at least 11 other squares each contain a letter that appears at least three times.

In the matrix, we will now colour all the squares which contain letters appearing at least three times in the same row with red, and all the squares which contain letters appearing at least three times in the same column with blue. By the above observation, each row contains at least 11 red squares, so the total number of red squares is at least 21×11. Similarly, each column contains at least 11 blue squares, so the total number of blue squares is at least 21×11. Since there are only 21×21 total entries in the matrix and

$$21 \times 21 < 21 \times 11 + 21 \times 11,$$

some square (or squares) is coloured both red and blue. Because the letter in this square appears in three different columns and three different rows, at least three girls and three boys solved the corresponding problem. Thus, we have found the problem(s) satisfying the desired property.

(148) *Problem.*   Using only the operations of addition, subtraction and the finding of reciprocals, show how to compute the product of two real numbers.

Solution.   Let a, b be any two real numbers. We will demonstrate how to compute the product ab using only the allowable operations of addition, subtraction, and reciprocation. First, we observe that if either $a = 0$ or $b = 0$, the answer is 0 and we are done. Thus, we may assume that neither a nor b is 0.

An observation that you may have made in the past turns out to be key to solving this problem, namely

$(a + b)^2 = a^2 + 2ab + b^2$ or, instead $2ab = (a + b)^2 - a^2 - b^2$.

Now, if we could find a way of producing squares and of taking half of a given value, we would be done. The second of these is easier to see, since

$$\frac{1}{2ab} + \frac{1}{2ab} = \frac{1}{ab},$$

and taking another reciprocal would yield ab. Thus, it remains to find a way of producing x^2 given x. If we ignore for the moment the worry about division by 0, then we observe the following:

$$\frac{1}{\dfrac{1}{x-1}-\dfrac{1}{x}}+x = \frac{1}{\dfrac{1}{x^2-x}}+x = x^2$$

and

$$\frac{1}{\dfrac{1}{x}-\dfrac{1}{x+1}}-x = \frac{1}{\dfrac{1}{x^2+x}}-x = x^2.$$

The first works as long as $x \neq 0$ and $x \neq 1$; the second works as long as $x \neq 0$ and $x \neq -1$. Therefore, at least one of them works as long as $x \neq 0$. But $0^2 = 0$. We can now define two functions, f and g, as follows: $f(x) = x^2$ and $g(x,y) = f(x+y) - f(x) - f(y)$, where, in computing $f(x) = x^2$, we use the approach that if $x = 0$ then $f(x) = 0$, and if $x \neq 0$ we compute $f(x)$ using either of the formulas displayed above to get x (at most one of them will involve division by 0 and will be unusable). Then the function h defined as:

$$h(x,y) = \frac{1}{g(x,y)} + \frac{1}{g(x,y)}$$

satisfies the property that it is computed using only addition, subtraction, and reciprocation, and furthermore we have shown above that $h(a,b) = ab$.

(149) *Problem.* Suppose you wish to play a card-guessing trick, working together with an assistant, on some volunteer victim. The trick is to be played as follows: Have a volunteer deal out to your assistant any five cards from an ordinary 52-card deck of playing cards. The assistant looks at the cards, gives one of them back to the volunteer who hides it from you, and then displays the other four cards face-up in a row on a table for you to look at. With just those four cards (and their order) to look at, you announce to the volunteer (and any other spectators) the hidden card! (That is, you specify both its rank and its suit.)

Devise a method which you and your assistant could use to make this trick work, no matter which five cards you are dealt. Remember, you are only allowed to use the four face-up cards and the order in which they are displayed.

Solution. Observe first that since there are five cards dealt out and only four suits, we may be assured that among the five cards dealt out, two will be in the same suit. By displaying one of those two cards as the first card then determines the suit.

Since the hidden card must be different from the first card displayed, this leaves only twelve possible values for the hidden card. There are still three other cards to be displayed in some order. Unfortunately, there are only 3! = 6 possible ways to order those three cards. Thus, we must think of something else. The one thing we have not discussed is which of the two cards that are in the same suit should be displayed and which hidden. Perhaps there is some way to reduce the number of possibilities down to six or less. As it turns out, there is a way to do this.

Think of the cards in each suit as a cycle:

$$A \to 2 \to 3 \to \cdots \to 10 \to J \to Q \to K \to A.$$

Then we observe that no matter which two cards in the suit are selected, one of the two gaps between the cards has five or fewer cards in it; that is, one of the two cards is at most six places to the right of the other, in the above cycle. This is sufficient for us to develop a workable method. Let us agree to always choose the smaller gap and display the left-most card of the two cards surrounding the gap, and hide the other one. (If there are three or more cards in that suit, we just simply use any two of them for the above purpose.)

Having decided on which card to display, we now need to decide how to interpret the six possible orders. (It is actually up to you as to how to do this, since we know the trick can be done.) We will agree to use the following linear order of all fifty-two cards: the suits are ordered alphabetically (Clubs, Diamonds, Hearts, and Spades), and within each suit we use the numerical order with A low and J-Q-K high (in that order). That way the three cards still to be displayed in some order already have a natural linear order inherited from the above. Since 1-2-3 also has a natural order, let us use it to describe how to proceed:

$(1, 2, 3) \implies$ next card right of displayed card in cycle

$(1, 3, 2) \implies$ two cards to the right of displayed card in cycle

$(2, 1, 3) \implies$ three cards to the right of displayed card in cycle

$(2, 3, 1) \implies$ four cards to the right of displayed card in cycle

$(3, 1, 2) \implies$ five cards to the right of displayed card in cycle

$(3, 2, 1) \implies$ six cards to the right of displayed card in cycle

(150) *Problem.*   Alphonse and Beryl are playing a game, starting with a pack of seven cards. Alphonse begins by discarding at least one but not more than half of the cards in the pack. He then passes the remaining cards in the pack to Beryl. Beryl continues the game by discarding at least one but not more than half of the remaining cards in the pack. The game continues in this way with the pack being passed back and forth between the two players. The loser is the player who, at the beginning of his or her turn, receives only one card. Show, with justification, that there is always a winning strategy for Beryl.

Alphonse and Beryl now play a game with the same rules as above, except this time they start with a pack of 52 cards. Alphonse goes first again. As in the game above, a player on his or her turn must discard at least one and not more than half of the remaining cards from the pack. Is there a strategy that Alphonse can use to be guaranteed that he will win? (Provide justification for your answer.)

Solution.   Alphonse starts with 7 cards, and so can remove 1, 2, or 3 cards, passing 6, 5, or 4 cards to Beryl. Beryl should remove 3, 2, or 1 cards, respectively, leaving 3 cards only, and pass these 3 cards back to Alphonse. Alphonse now is forced to remove 1 card only, and pass 2 back to Beryl. Beryl removes 1 card (her only option) and passes 1 back to Alphonse, who thus loses. Therefore, Beryl is guaranteed to win.

Alphonse removes 21 cards from original 52, and passes 31 cards to Beryl. If Beryl removes b_1 cards with $1 \le b_1 \le 15$, then Alphonse removes $16 - b_1$ cards to reduce the pack to 15 cards. [This is always a legal move, since $2(16 - b_1) = 32 - 2b_1 \le 31 - b_1$ so $16 - b_1$ is never more than half of the pack.] If Beryl removes b_2 cards with $1 \le b_2 \le 7$, then Alphonse removes $8 - b_2$ to reduce the pack to 7 cards. [This move is always legal, by a similar argument.] Since Beryl now has 7 cards, Alphonse can adopt Beryl's strategy from the first paragraph. Thus, Alphonse has a winning strategy.

What would be your strategy in playing this game with a deck of n cards? What value(s) of n would give Alphonse (the first player) a winning strategy, and what value(s) of n would give Beryl (the second player) a winning strategy?

(151) *Problem.*   Bruce is down at Hunk's Gym riding the bicycle when a drop-dead gorgeous woman walks in. Bruce, always on the lookout, goes right over to her and strikes up a conversation. They quickly discover they grew up on the same street.

Bruce is a pretty smart guy; he teaches math at the local college. He makes a deal with his new acquaintance that if he can ask her five yes-or-no questions and guess her address with her answers, she'll go out to dinner with him the next evening.

Bruce knows that all of the addresses on Elm Street are three-digit numbers. His first four questions are:

1. "Is your address a perfect square?"
2. "Is it greater than 600?"
3. "Is it a perfect cube?"
4. "Is it an even number?"

When he hears his dream date's answers to the first four questions, Bruce gets real excited. "Now, once I know whether your address contains a 6 or not, I'll know your address."

"Yes, it does contain a 6," the woman answers.

Bruce, grinning from ear to ear, replies, "Aha! I'll pick you up tomorrow at seven."

The next night at half past seven, Bruce shows up alone at Hunk's. "What happened?" asked his trainer, who had overheard Bruce's little game the night before.

"The address turned out to be a vacant lot."

"Sorry," the trainer comments, though he is not surprised. He would have suspected that the woman had not been entirely truthful in her responses to Bruce's questions.

In fact, as it turned out, she had lied in her answer to every question.

Where did Bruce go to pick up his date, and what is her real address?

Solution.   From Bruce's remark after question 4, it is clear that he had it narrowed down to a choice of 2 possibilities. So, let us examine the possibilities associated with different combinations of answers.

If the answer to questions 1 and 3 are both NO, then no matter how questions 2 and 4 are answered, there are a lot of acceptable addresses, certainly more than the two required. Thus, she must have answered YES to either or both of questions 1 and 3. On the other hand, if she answered YES to both questions 1 and 3, there is only one possible answer, namely 729, and no further questions would be necessary. So we may eliminate that possibility also. Therefore, she answered YES to exactly one of questions 1 and 3. Let us first suppose she answered YES to question 1 (and NO to question 3). Then the even perfect squares greater than 600 are 676, 784, and 900, which is one too many. Those less than 600 are 100, 144, 196, 256, 324, 400, 484, and 576, which is again too many. The odd perfect squares greater than 600 are 625, 729, 841, and 961; even though we can eliminate 729 (since it is also a perfect cube), we have one too many. The odd perfect squares less than 600 are 121, 169, 225, 289, 361, 441, and 529, again too many. Therefore, we may conclude that her answer to question 1 was NO, which means that she answered YES to question 3.

Then we know that the address is one of 125, 216, 343, 512. Clearly, the answer to question 2 was NO. If the answer to question 4 was NO, then we have to choose between 125 and 343. Knowing about the presence or absence of the digit 6 in the address is not going to help out here. Thus, this possibility must be rejected. Therefore, her answer to question 4 must have been YES, which leaves us with a choice between 216 and 512. Since she claimed there was a 6 in her address, Bruce went to 216 Elm Street.

Her answers to the first 4 questions were, in order, NO-NO-YES-YES. Since they were all lies, to find her real address we would use YES-YES-NO-NO. From above we see that we must choose between 625, 841, and 961. Since she claimed there was a digit 6 in her address and she lied, her real address must be 841 Elm Street.

(152) *Problem.* There are 2003 marbles in a pile. The pile is divided into two smaller piles, the number of marbles in each of the two smaller piles is counted, and the product of these two numbers is written down.

A pile containing at least two marbles is then selected, divided into two smaller piles, the number of marbles in each of these two new smaller piles is counted, and the product is written down. This procedure is continued until every pile contains exactly one marble. Find the maximum sum of the 2002 products written down.

Solution. Pretend that the marbles have hands and that every two marbles shake hands when they are separated. When a pile of marbles is divided into two smaller piles, the product generated by the problem statement simply counts the number of handshakes that are caused by that separation.

In the end every two marbles shake hands exactly once. Since there are 2003 marbles, the total number of handshakes is $\frac{1}{2} \cdot 2003 \cdot 2002 = 2005003$. This will be the sum, regardless of what sequence of divisions is used.

For those who do not know how to count handshakes among a group, here is an explanation. If there are n persons in the group, then each person shakes hands with $n - 1$ other persons. This yields $n(n - 1)$ handshakes. However, each such handshake has been counted twice, once for each party to the handshake. Thus, to get the actual number of handshakes, we need to divide this count by 2, yielding a total of $\frac{1}{2}n(n - 1)$ handshakes.

(153) *Problem.* A square array of dots with 10 rows and 10 columns is given. Each dot is coloured either blue or red. Whenever two dots of the same colour are adjacent in the same row or column, they are joined by a line segment of the same colour as the dots. If they are adjacent, but of different colour, they are joined by a green line segment.

In total there are 52 red dots, including 18 on the edges, 2 of which are at the corner of the array. There are 98 green line segments.

Determine all the possible values for the number of blue line segments.

Solution. In each row and column, there are 9 line segments, giving a total of 180 line segments (90 for the rows and 90 for the columns). Each corner dot has the ends of 2 line segments attached to it, each side dot (other than a corner) has the ends of 3 line segments attached to it, and each interior dot has the ends of 4 line segments attached to it. Thus, coming out of the red dots, there are:

$$2 \cdot 2 + (18 - 2) \cdot 3 + (52 - 18) \cdot 4 = 188$$

ends of line segments. Of these, 98 are accounted for by the green line segments, leaving 90 for the red segments. Since the red line segments account for 2 dots at a time, this means we have accounted for 45 red line segments. This leaves us with $180 - 98 - 45 = 37$ blue line segments as the only possible solution.

(154) *Problem.* Two young boys discovered in the attic of a neighbour-hood magician three abandoned Transmogrifiers.

The Dogmatizer would change one dog into one cat and one rat.
The Categorizer would change one cat into two dogs and one rat.
The Rationalizer would change one rat into three dogs and one cat.
Each machine would also work in reverse. Starting with magician's cat, the boys tried to obtain

(a) some rats, but without any dogs or cats;
(b) some dogs, but without any cats or rats;
(c) more cats, but without any dogs or rats.

For each task, either prove that it is impossible or find a sequence of transmogrifications which will result in the minimum number of animals of the specified type.

Solution. Let Δd, Δc, and Δr represent the changes in the numbers of dogs, cats, and rats, respectively, after i applications of the Dogmatizer, j applications of the Categorizer, and k applications of the Rationalizer. Then we obtain the system of equations:

$$-i + 2j + 3k = \Delta d \tag{1}$$
$$i - j + k = \Delta c \tag{2}$$
$$i + j - k = \Delta r. \tag{3}$$

Adding (2) and (3) yields

$$2i = \Delta c + \Delta r, \tag{4}$$

which simplifies to $i = \frac{1}{2}(\Delta c + \Delta r)$. Subtracting (2) from (3) yields

$$2j - 2k = \Delta r - \Delta c. \tag{5}$$

Adding (1) and (3) yields $3j + 2k = \Delta d + \Delta r$, which, when added to (5), gives us $5j = \Delta d - \Delta c + 2\Delta r$.
Combining this and (4) with (1), (2), or (3), we obtain $10k = 2\Delta d + 3\Delta c - \Delta r$, which means our system has solution:

$$i = \frac{1}{2}\Delta c + \frac{1}{2}\Delta r,$$

$$j = \frac{1}{5}\Delta d - \frac{1}{5}\Delta c + \frac{2}{5}\Delta r,$$

$$k = \frac{1}{5}\Delta d + \frac{3}{10}\Delta c - \frac{1}{10}\Delta r.$$

This system could also have been solved using matrix algebra: the system can be first rewritten as the matrix equation:

$$\begin{bmatrix} -1 & 2 & 3 \\ 1 & -1 & 1 \\ 1 & 1 & -1 \end{bmatrix} \cdot \begin{bmatrix} i \\ j \\ k \end{bmatrix} = \begin{bmatrix} \Delta d \\ \Delta c \\ \Delta r \end{bmatrix}. \tag{6}$$

The columns of the matrix refer to the three Transmogrifiers, while the rows refer to the three animals. The inverse of this matrix is found to be

$$\begin{bmatrix} 0 & \dfrac{1}{2} & \dfrac{1}{2} \\ \dfrac{1}{5} & -\dfrac{1}{5} & \dfrac{2}{5} \\ \dfrac{1}{5} & \dfrac{3}{10} & -\dfrac{1}{10} \end{bmatrix}.$$

Thus, for any given values of Δd, Δc, and Δr, we obtain the number of applications of each of the transmogrifiers by multiplying equation (6) by this inverse to get

$$\begin{bmatrix} i \\ j \\ k \end{bmatrix} = \begin{bmatrix} 0 & \dfrac{1}{2} & \dfrac{1}{2} \\ \dfrac{1}{5} & -\dfrac{1}{5} & \dfrac{2}{5} \\ \dfrac{1}{5} & \dfrac{3}{10} & -\dfrac{1}{10} \end{bmatrix} \cdot \begin{bmatrix} \Delta d \\ \Delta c \\ \Delta r \end{bmatrix}.$$

(a) In this case we want no dogs at the end ($\Delta d = 0$), no cats either ($\Delta c = -1$, since we started with one cat), and some integer number n of rats ($\Delta r = n$). Assigning these values to our solution system, we have

$$i = \frac{n-1}{2}, \qquad j = \frac{2n+1}{5}, \qquad \text{and} \qquad k = -\frac{n+3}{10}.$$

The smallest positive integer value of n which makes all of i, j, and k integral is $n = 7$, which gives $i = 3$, $j = 3$, and $k = -1$. This means that we need to apply the Dogmatizer and Categorizer 3 times each and the Rationalizer 1 time in reverse to end up with 7 rats. Note that since we only have a cat to start, we must begin the process by applying the Categorizer.

(b) In this case we want no cats at the end ($\Delta c = -1$), no rats either ($\Delta r = 0$), and some integer number n of dogs ($\Delta d = n$). This yields the value $i = -\frac{1}{2}$, a non-integer value, which is impossible.

(c) In this last case we want no dogs at the end ($\Delta d = 0$), no rats either ($\Delta r = 0$), and some integer number n of cats ($\Delta c = n - 1$). This yields

$$i = \frac{n-1}{2}, \qquad j = -\frac{n-1}{5}, \qquad \text{and} \qquad k = \frac{3(n-1)}{10}.$$

The smallest positive integer value of n which makes all of i, j, and k integral is $n = 11$, which gives $i = 5$, $j = -2$, and $k = 3$. This means that we need to apply the Dogmatizer 5 times, the Rationalizer 3 times, and the Categorizer 2 times in reverse to end up with 11 cats (10 more than we started with). However, since we only have a cat to start, we must begin the process by applying the Categorizer. This means that we need at the end to apply the Categorizer 3 times in reverse, in order to have a net of 2 times in reverse.

(155) *Problem.* Prove that in any set of ten different 2–digit numbers one can select two disjoint subsets such that the sum of the numbers in each of the subsets is the same.

Solution. Any set of 10 elements has $2^{10} - 1 = 1023$ distinct non-empty subsets. Let S be any set of ten different 2–digit numbers. Since the smallest possible sum of numbers taken from S is 10 (if 10 is in S), and the largest possible sum is 945 (if $S = \{90, 91, \ldots, 99\}$), this leaves a total of 936 different possible sums. Since $1023 > 936$, there must be at least two distinct non-empty subsets of S which have the same sum. Let A_1 and A_2 be two such subsets of S. If they are disjoint, then we are done. If (instead) they have elements in common, then we simply remove those elements from both sets, and we are again done.

(156) *Problem.* "The product of the ages of my three children is less than 100," said Bill, "but even if I told you the exact product, and even told you the sum of their ages, you still could not figure out each child's age."

"I would have trouble if different ages are very close," said John as he looked at the children, "but tell me the product anyway."

Bill told him. When Bill's children came into the room, John confidently told each child his age.

Given that you can now determine their ages, what are the three children's ages?

Solution. Since the knowledge of both their product and sum does not determine the ages uniquely, the three ages must form one of a set of triples all of which have the same product and the same sum. Given that the product is less than 100, one can check (see below) that the only possibilities are in the following table:

Product	Sum	Set of triples	
36	13	$\{2,2,9\}$	$\{1,6,8\}$
40	14	$\{2,2,10\}$	$\{1,5,8\}$
72	14	$\{3,3,8\}$	$\{2,6,6\}$
90	16	$\{2,5,9\}$	$\{3,3,10\}$
90	20	$\{2,3,15\}^*$	$\{1,9,10\}^*$
96	21	$\{1,8,12\}$	$\{2,3,16\}^*$

Since John has trouble with ages which are close, yet was confidently able to associate an age with each child, the correct triple cannot be one with consecutive ages. This eliminates the starred triples.

We are also told that we can now tell the ages (even though we do not get to see the children or know the product that Bill gave). Had the given product been 36, for example, John could have chosen the correct set by seeing the children, but we could not. This eliminates the first four rows of the table.

Thus, the given product must be 96, for which there is only one acceptable triple remaining, namely $\{1,8,12\}$, which gives us the correct ages.

Let us now fill in some of the details missing from the preamble (above the table). Let the ages of the children be a, b, and c, where $a \leq b \leq c$, and let the product of the ages be P. That is, $P = abc$. We are seeking at least two sets of three ages such that the product P and the sum are the same. We will now eliminate some special cases.

Case (i): P is a prime.
We may rule out this case, since we could only have one set of ages, namely $(a,b,c) = (1,1,P)$.

Case (ii): $P = p^k$, where p is a prime, and $k \geq 2$.
The only possible ages in this case are powers of p. In order for the sum of two such powers of p to be equal, they must be the same powers (simply think of representing numbers in base p). Thus, we may rule out this case also.

Case (iii): $P = pq$, where p and q are primes and $p < q$.
The only possible ages in this case are $(a, b, c) = (1, 1, pq)$ and $(a, b, c) = (1, p, q)$. The respective sums are $pq + 2$ and $p + q + 1$. Setting these sums equal, we get

$$(p - 1)(q - 1) = 0,$$

which is impossible for primes p and q.

Case (iv): $P = p^2q$, where p and q are distinct primes.
The ages then are one of the following (not necessarily ordered) triples:

    (a) $(1, 1, p^2q)$ with sum $S_1 = p^2q + 2$,
    (b) $(1, q, p^2)$ with sum $S_2 = p^2 + q + 1$,
    (c) $(1, p, pq)$ with sum $S_3 = pq + p + 1$, or
    (d) (p, p, q) with sum $S_4 = 2p + q$.

If $S_1 = S_2$, then we get $(p^2 - 1)(q - 1) = 0$, which is impossible. If $S_1 = S_3$, then we have $p^2q - pq - p + 1 = 0$, or $p(pq - q - 1) + 1 = 0$, which implies that p must must divide 1, which is impossible. If $S_1 = S_4$, then we get $(p - 1)(pq + q - 2) = 0$, which forces $pq + q = 2$, again impossible for primes p and q. If $S_2 = S_3$, then we get $(p - 1)(q - p) = 0$, impossible since p and q are distinct. If $S_2 = S_4$, then we get $(p - 1)(q - 1) = 0$, impossible. Finally, if $S_3 = S_4$, then p must be 1, impossible. Thus, we may eliminate this case also.

Case (v): $P = pqr$, where p, q, and r are distinct primes.
The ages then are one of the following (not necessarily ordered) triples:

    (a) $(1, 1, pqr)$ with sum $S_1 = pqr + 2$,
    (b) $(1, p, qr)$ with sum $S_2 = qr + p + 1$,
    (c) $(1, q, pr)$ with sum $S_3 = pr + q + 1$,
    (d) $(1, r, pq)$ with sum $S_4 = pq + r + 1$, or
    (e) (p, q, r) with sum $S_4 = p + q + r$.

Since (b), (c), and (d) are essentially the same, we need only treat one of them in comparing with the others. We will treat (b). If $S_1 = S_2$, then we get $(p - 1)(qr - 1) = 0$, which is impossible. If $S_1 = S_5$, then we have $pqr - p - q - r + 2 = 0$, or $(p - 1)(q - 1) + (pq - 1)(r - 1) = 0$, which is impossible since all terms are positive. If $S_2 = S_5$, then we get $(q - 1)(r - 1) = 0$, again impossible for primes q and r. We still need to compare two of (b), (c), and (d).

If $S_2 = S_3$, then we get $(r - 1)(q - p) = 0$, impossible since p and q are distinct. Thus, we may eliminate this case also.

Since $P < 100$, this leaves us with

$$P \in \{24, 36, 40, 48, 54, 56, 60, 72, 80, 84, 88, 90, 96\}.$$

We must examine each in turn to see if there are two sets of ages with the same sum.

Case: $P = 24$.

Set of ages	Sum
$\{1, 1, 24\}$	26
$\{1, 2, 12\}$	15
$\{1, 3, 8\}$	12
$\{1, 4, 6\}$	11
$\{2, 2, 6\}$	10
$\{2, 3, 4\}$	9

Since the sums are distinct, we may eliminate $P = 24$.

Case: $P = 36$.

Set of ages	Sum
$\{1, 1, 36\}$	38
$\{1, 2, 18\}$	21
$\{1, 3, 12\}$	16
$\{1, 4, 9\}$	14
$\{1, 6, 6\}$	13
$\{2, 2, 9\}$	13
$\{2, 3, 6\}$	11

Since there are two sets of ages which sum to 13, we keep $P = 36$ as a candidate.

Case: $P = 40$.

Set of ages	Sum
$\{1, 1, 40\}$	42
$\{1, 2, 20\}$	23
$\{1, 4, 10\}$	15
$\{1, 5, 8\}$	14
$\{2, 2, 10\}$	14
$\{2, 4, 5\}$	11

Since there are two sets of ages which sum to 14, we keep $P = 40$ as a candidate.

Case: $P = 48$.

Set of ages	Sum
$\{1, 1, 48\}$	50
$\{1, 2, 24\}$	27
$\{1, 3, 16\}$	20
$\{1, 4, 12\}$	17
$\{1, 6, 8\}$	15
$\{2, 2, 12\}$	16
$\{2, 3, 8\}$	13
$\{2, 4, 6\}$	12
$\{3, 4, 4\}$	11

Since the sums are distinct, we may eliminate $P = 48$.

Case: $P = 54$.

Set of ages	Sum
$\{1, 1, 54\}$	56
$\{1, 2, 27\}$	30
$\{1, 3, 18\}$	22
$\{1, 6, 9\}$	16
$\{2, 3, 9\}$	14
$\{3, 3, 6\}$	12

Since the sums are distinct, we may eliminate $P = 54$.

Case: $P = 56$.

Set of ages	Sum
$\{1, 1, 56\}$	58
$\{1, 2, 28\}$	31
$\{1, 4, 14\}$	19
$\{1, 7, 8\}$	16
$\{2, 2, 14\}$	18
$\{2, 4, 7\}$	13

Since the sums are distinct, we may eliminate $P = 56$.

Case: $P = 60$.

Set of ages	Sum
$\{1, 1, 60\}$	62
$\{1, 2, 30\}$	33
$\{1, 3, 20\}$	24
$\{1, 4, 15\}$	20
$\{1, 5, 12\}$	18
$\{1, 6, 10\}$	17
$\{2, 2, 15\}$	19
$\{2, 3, 10\}$	15
$\{2, 5, 6\}$	13
$\{3, 4, 5\}$	12

Since the sums are distinct, we may eliminate $P = 60$.

Case: $P = 72$.

Set of ages	Sum
$\{1, 1, 72\}$	74
$\{1, 2, 36\}$	39
$\{1, 3, 24\}$	28
$\{1, 4, 18\}$	23
$\{1, 6, 12\}$	19
$\{1, 8, 9\}$	18
$\{2, 2, 18\}$	22
$\{2, 3, 12\}$	17
$\{2, 4, 9\}$	15
$\{2, 6, 6\}$	14
$\{3, 3, 8\}$	14
$\{3, 4, 6\}$	13

Since there are two sets of ages which sum to 14, we keep $P = 72$ as a candidate.

Case: $P = 80$.

Set of ages	Sum
$\{1, 1, 80\}$	82
$\{1, 2, 40\}$	43
$\{1, 4, 20\}$	25
$\{1, 5, 16\}$	22
$\{1, 8, 10\}$	19
$\{2, 2, 20\}$	24
$\{2, 4, 10\}$	16
$\{2, 5, 8\}$	15
$\{4, 4, 5\}$	13

Since the sums are distinct, we may eliminate $P = 80$.

Case: $P = 84$.

Set of ages	Sum
$\{1, 1, 84\}$	86
$\{1, 2, 42\}$	45
$\{1, 3, 28\}$	32
$\{1, 4, 21\}$	26
$\{1, 6, 14\}$	21
$\{1, 7, 12\}$	20
$\{2, 2, 21\}$	25
$\{2, 3, 14\}$	19
$\{2, 6, 7\}$	15
$\{3, 4, 7\}$	14

Since the sums are distinct, we may eliminate $P = 84$.

Case: $P = 88$.

Set of ages	Sum
$\{1, 1, 88\}$	90
$\{1, 2, 44\}$	47
$\{1, 4, 22\}$	27
$\{1, 8, 11\}$	20
$\{2, 2, 22\}$	26
$\{2, 4, 11\}$	17

Since the sums are distinct, we may eliminate $P = 88$.

Case: $P = 90$.

Set of ages	Sum
$\{1, 1, 90\}$	92
$\{1, 2, 45\}$	48
$\{1, 3, 30\}$	34
$\{1, 5, 18\}$	24
$\{1, 6, 15\}$	22
$\{1, 9, 10\}$	20
$\{2, 3, 15\}$	20
$\{2, 5, 9\}$	16
$\{3, 3, 10\}$	16
$\{3, 5, 6\}$	14

Since there are two sets of ages which sum to 20 and another two sets of ages which sum to 16, we keep $P = 90$ as a candidate.

Case: $P = 96$.

Set of ages	Sum
$\{1, 1, 96\}$	98
$\{1, 2, 48\}$	51
$\{1, 3, 32\}$	36
$\{1, 4, 24\}$	29
$\{1, 6, 16\}$	23
$\{1, 8, 12\}$	21
$\{2, 2, 24\}$	28
$\{2, 3, 16\}$	21
$\{2, 4, 12\}$	18
$\{2, 6, 8\}$	16
$\{3, 4, 8\}$	17
$\{4, 4, 6\}$	14

Since there are two sets of ages which sum to 21, we keep $P = 96$ as a candidate.

We have determined that the only candidates for P are 36, 40, 72, 90 (twice), and 96, which is where the solution began.

(157) *Problem.*   You are given a box containing balls of three different colours, along with a supply of each colour of ball on the side. Randomly choose two balls from the box. If the balls are of the same colour, return them to the box; if they are of different colours,

replace them by two balls of the third colour. The goal is to end up with all the balls in the box having the same colour.

(a) Can the goal be achieved if we start with 3 green, 4 yellow, and 5 red balls in the box?

(b) Under what initial conditions can the goal be achieved?

Solution. To achieve the goal it is necessary to have the number of balls of two different colours the same at some point. Then by removing one of each, we will be able to replace all of them with the third colour.

Let g_n, y_n, and r_n be the number of green, yellow, and red balls (respectively) in the box after n rounds, where g_0, y_0, and r_0 are the initial numbers. (We agree to ignore any round in which two balls of the same colour are chosen.) In each round one of the three values increases by 2 while the other two values decrease by 1. Thus, we must have

$$|g_{n+1}-y_{n+1}| = \begin{cases} |g_n - y_n| & \text{if a green and yellow ball were chosen,} \\ 3 & \text{otherwise.} \end{cases}$$

We can make a similar observation about the other differences. This means that the only way to get two of the numbers equal is to start by having the difference between them a multiple of 3. That is, as long as the number of balls of the three different colours do NOT give three different remainders on division by 3 we can achieve the goal; if they give three different remainders, we cannot achieve the goal.

In the given initial configuration the remainders on division by 3 are 0, 1, and 2, which means the goal is unattainable.

(158) *Problem.* In a certain multiple-choice test, one of the questions was illegible, but the choice of answers was clearly printed. Determine the true answer(s) from the given choices:

    (a) All of those below.

    (b) None of those below

    (c) All of the above.

    (d) One of the above.

    (e) None of the above.

    (f) None of the above.

Solution. If we assume that either (a) or (f) is true, then (e) gives us a contradiction. Therefore, (a) and (f) are both false. Since (a) is false, so is (c). If we now assume that (b) is true, then (d) would also be true, which contradicts (b). Therefore, (b) must be false. Now we conclude that (d) is also false since (a), (b), and (c) are all false. That leaves (e) as the only candidate to be true. Since (a), (b), (c), and (d) are all false, we see that (e) is indeed true.

(159) *Problem.*   At midnight a truck starts from city A and goes to city B; at 2:00 am a car starts along the same route from city B to city A. They pass each other at 4:00 am. The car arrives at its destination 40 minutes later than the truck. Having their respective business, they start for home and pass each other on the road at 2:00 pm. Finally, they both arrive home at the same time. At what time did they arrive home? (Assume that both vehicles travel at constant speeds.)

Solution.   Let us draw a graph of the time versus the distance with respect to city A and city B:

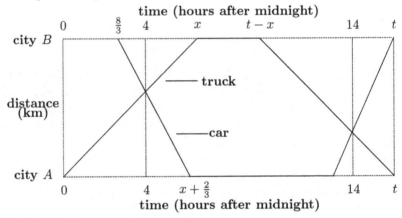

Let v_c be the (constant) speed (in km/h) of the car and let v_t be that of the truck. Let x be the time (in hours after midnight) that the truck reaches city B. Then the car reaches city A at $x + 2/3$ hours after midnight. Then the distance (in kms) from city A to the point where they first meet can be measured in two ways as:

$$\left(x + \frac{2}{3} - 4 \right) = 4v_t.$$

Similarly, the distance from this point to city B is given by

$$\left(4 - \frac{8}{3} \right) v_c = (x - 4)v_t.$$

Dividing the first equation by the second and simplifying yields $(3x - 4)(x - 6) = 0$. Since $x > 4$ (because they meet at $x = 4$

and at that point neither vehicle has arrived at its destination), we see that $x = 6$. Using this in either of the two displayed equations above gives us $v_c/v_t = 3/2$. Now, the distance of the second point where they pass each other from city B can be computed in two different ways as:

$$(t - 14)v_c = \big(14 - (t - 6)\big)v_t.$$

Using $v_c/v_t = 3/2$ in this equation yields $t = 16.4$. Thus, they both arrive home at 4:24 pm.

(160) *Problem.* Each of two rooms contained either a treasure or a tiger. A sign on the door of the first room read: "In this room there is a treasure, and in the other room there is a tiger." A sign on the door of the second room read: "In one of these rooms there is a treasure, and in one of these rooms there is a tiger." One of the signs was true, but the other was false. It was possible that both rooms contained treasures or both rooms contained tigers. Which room should be chosen in order to get a treasure?

Solution. If the sign on the first room is true, so is the sign on the second, which is impossible. Thus, the sign on the first room is false, and the sign on the second room is true. Let us suppose that there is no treasure in the second room. Then the second room must clearly contain a tiger (because of the first sentence in the problem statement), and the first room must clearly contain a treasure because of the (true) sign on the second room. But this would imply that the sign on the first room is also true, a case which has already been ruled out. Thus, there must be a treasure in the second room.

(161) *Problem.* A railway line is divided into 10 sections by the stations A, B, C, D, E, F, G, H, I, J, and K (in that order). The distance between A and K is 56 km. A trip along two successive sections never exceeds 12 km. A trip along three successive sections is at least 17 km. What is the distance between B and G? (See diagram below.)

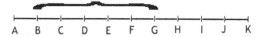

A   B   C   D   E   F   G   H   I   J   K

Solution. Since any single section can be considered as the difference of some three successive sections and a subset of two successive sections, we see that any single section must be at least

$17 - 12 = 5$ km in length. Next note that section JK is the entire line minus the 3 sets of three successive sections: AD, DG, and GJ.

Since these three sets of three successive sections must total at least 51 km, we see that section JK can be at most 5 km in length, which (in view of our first argument) implies that section JK is exactly 5 km in length. By symmetry we see that AB is also exactly 5 km in length. We can also lay out three sets of successive sections so as to isolate section DE or GH; then the same argument as above can be used to deduce that each of them is exactly 5 km in length. Since the three sets of two successive sections remaining, namely BD, EG, and HJ, can sum to at most $3 \times 12 = 36$ km, and at the same time must cover the remaining distance: $56 - 4 \times 5 = 36$, we see that each of these three sets of two successive sections must be exactly 12 km in length. Thus, the total length from B to G is exactly $12 + 5 + 12 = 29$ km.

(162) *Problem.* Alice, Betty, and Carol took the same series of examinations. For each examination there was one mark of x, one mark of y, and one mark of z, where x, y, z are distinct positive integers. The total of the marks obtained by each of the girls was: Alice – 20; Betty – 10; Carol – 9. If Betty placed first in the algebra examination, who placed second in the geometry examination?

Solution. Let us assume that $x > y > z > 0$. Since there were three of each mark awarded to the girls we see that

$$3(x + y + z) = 20 + 10 + 9 = 39,$$

implying that $x + y + z = 13$. We are told that Betty was first in algebra. Thus, Betty has a score of x plus two other scores which total 10. This tells us that $x \le 8$. Also note that a score of 20 can only be obtained by having at least one score of 7 or better. These two facts force us to conclude that $x = 7$ or $x = 8$. In either case Betty (who has one score of x) must have a score of 1, which implies that $z = 1$. Using this we see that $x = 7$ further implies that $y = 2$, which is impossible since $x + y + z = 13$. Therefore, $x = 8$ and $z = 1$, which implies that $y = 4$.

Therefore, Alice (with a score of 20) must have two first place finishes (in geometry and the other unnamed subject) and a second place (in algebra, since Betty came first in algebra), and Betty has a first (in algebra) and two third place finishes, leaving Carol with

two second places and a third place. Carol's second place finishes
are in geometry and the unnamed subject (since Alice was second
in algebra). Thus, Carol finished second in geometry.

(163) *Problem.*   Five persons are seated at a round table. Each one
remarks in his turn "My two neighbours, to the right and to the
left, are both liars". We know that liars always lie, and non-liars
always tell the truth.

Moreover, everyone at the table knows for certain who is a liar and
who is a non-liar. How many liars are at the table?

Solution.   It is clear that given any two persons seated beside each
other at least one must be a liar. This means that there must be
at least three liars at the table (if there were only two liars there
is a pair of adjacent seats occupied by two non-liars).

Next we observe that we cannot have three liars in consecutive
seats, for then the one in the middle would have told the truth.
From this it follows that we must have at least two non-liars at the
table, which means that we have at most three liars at the table.

Since we have put forward arguments for the number of liars to be
at least three, and also for the number to be at most three, we can
conclude that the number of liars at the table is exactly three.

Extension: How many liars would there be if there were n seats
occupied at the table? Is the number of liars always determined
completely by knowing the number of seats?

(164) *Problem.*   Alf, Bert, and Charlie decide to see if any one of them
is psychic. So they take four aces out of an ordinary deck of cards,
shuffle them thoroughly, and place three of them face down on a
table. Each person writes down his guess as to the identity of the
three cards. When they turned the aces over they discovered that
somebody made a correct guess for each card, that nobody made
a correct guess for all three cards, and that no two of them ended
up with the same number of correct guesses. What are the three
aces, and what position are they in?

	First Card	*Second Card*	*Third Card*
Alf	Spade	Club	Heart
Bert	Club	Diamond	Heart
Charlie	Club	Spade	Diamond

Solution.   Clearly, for each person there are none, one, two, or
three possible correct guesses. We are told that no one made three

correct guesses, and that no two of them made the same number of correct guesses. Therefore, Alf, Bert, and Charlie made in some order none, one, and two correct guesses.

Suppose first that Alf is the one who made no correct guesses. Then the first card is the Ace of Clubs and the third card is the Ace of Diamonds (since each card was correctly guessed). This forces the second card to be the Ace of Spades, which implies that Charlie made three correct guesses, which did NOT happen. Thus, Alf is NOT the one who made no correct guesses.

Suppose next that Charlie is the one who made no correct guesses. Then the first card is the Ace of Spades and the third card is the Ace of Hearts (again because each card was correctly guessed). This means that the second card must be either the Ace of Diamonds or the Ace of Clubs. If it was the Ace of Clubs then Alf would have made three correct guesses, so that possibility is eliminated. Thus, the second card must be the Ace of Diamonds. But in this case both Alf and Bert have two correct guesses, which is also impossible. Thus, Charlie canNOT be the one who made no correct guesses.

Therefore, Bert is the one who made no correct guesses. This means that the first card is the Ace of Spades and the third card is the Ace of Diamonds. This forces the second card to be the Ace of Clubs. This solution shows that Alf made two correct guesses, Bert made no correct guesses, and Charlie made one correct guess.

(165) *Problem.*   At noon, Jill, who is in training for a bicycle race, left Abbey to ride to Bakersfield and back again. It is 26 miles each way. She did the return journey without stopping and maintained a uniform speed throughout.

Some time later, Jack. trying out his new car, left Bakersfield and drove – also maintaining a uniform speed – to Abbey and back again. Jack passed Jill on the latter's outward journey 7.5 miles from Bakersfield and passed her again on her return journey 5.5 miles from Bakersfield. Jack finished the return journey at 3:20 pm.

What time was it when Jill was back at Abbey?

Solution.   Between their two meetings Jill covers 7.5 miles into Bakersfield and 5.5 miles back towards Abbey, a total of 13 miles, at a uniform speed of s miles per hour. During the same interval of time Jack covers 18.5 miles into Abbey ($26 - 7.5 = 18.5$) and

20.5 miles back toward Bakersfield $(26 - 5.5 = 20.5)$, a total of 39 miles, at a uniform speed of r miles per hour. Since these two times are equal, we have

$$\frac{13}{s} = \frac{39}{r},$$

from which we have $r = 3s$.

At the time of their second meeting, Jill had been pedalling for $(26 + 5.5)/s = 31.5/5$ hours. Also, the time from this meeting until Jack returns to Bakersfield at 3:20 pm is $5.5/r = 5.5/3s$ hours. Since the sum of these two times equals the difference between the time of Jack's return to Bakersfield and Jill's departure from Abbey, it follows that

$$\frac{31.5}{s} + \frac{5.5}{3s} = \frac{10}{3},$$
$$94.5 + 5.5 = 10s,$$

whence, $s = 10$ miles per hour. This means that Jill covered the 52 miles in 5.2 hours to arrive back at Abbey at 5:12 pm.

(166) *Problem.* Suppose you are visiting a forest in which every inhabitant is either a knight (who always tells the truth) or a knave (who always lies).

In addition, some of the inhabitants are werewolves and have the annoying habit of sometimes turning into wolves at night and devouring people. A werewolf can be either a knight or a knave.

You are interviewing three inhabitants — A, B, and C — and it is known that exactly one of them is a werewolf. They make the following statements:

A : C is a werewolf.

B : I am not a werewolf.

C : At least two of us are knaves.

Our problem has two parts:

1. Is the werewolf a knight or a knave?

2. If you have to take one of them as a travelling companion, and it is more important that he not be a werewolf than that he not be a knave, which one would you pick?

Solution. Clearly C must be a knight or a knave. Let us first assume that C is a knight. Then there must be at least two knaves, whence both A and B are knaves. Since B claims he is NOT a werewolf, he must indeed be the werewolf. Thus, if C is a knight, the werewolf is a knave. Next let us assume that C is a knave. Then C's statement is false, which means that C is the only knave. Since A is then a knight, we see that C is the werewolf. Again, the werewolf is a knave. Thus, in either case the werewolf is a knave, which answers the first question. Also the above analysis shows that either B or C is the werewolf, which means that you should take A as a travelling companion.

(167) *Problem.*   "Grandpa," said young Evelyn, "had such a bright idea the other day. He said he wanted to find out how greedy we were."

"What happened?" I asked.

"You tell him, Charles," said Evelyn.

"Well," said Charles, "he produced a big box of chocolates and gave it to Evelyn"

"No, he gave it to Kitty," corrected Evelyn.

"Anyway," said Charles, "he gave it to one of them.   Then he explained that all of us could eat as many chocolates as we liked. At the end of the day, Kitty would report the number of chocolates eaten. Then each of us would receive, from Grandpa, one penny for each chocolate eaten by somebody else, less three cents for each chocolate eaten by himself (or herself)."

"What a cunning scheme!" said I. "I'll bet it didn't cost Grandpa very much."

"Didn't it just!" said Clare. "It cost him all of $4.69."

"Whoever ate all the chocolates?" I asked.

"Edna ate the most," said Charles. "Then came David. I forget about the rest of us. But I remember that no two of us ate the same number, and that each of us ate at least one."

How many children in all ate chocolates, and how many did Edna eat?

Solution.   Grandpa pays out, for each chocolate eaten, as many cents as there are children less one cent (not paid to the child who eats the chocolate), less a further three cents (deducted from the sum which the eater of the chocolate would otherwise receive).

That is, if there are n children and c chocolates eaten,

$$(n - 4)c = 469.$$

The only factors of 469 are 1, 7, 67, and 469, so that $n - 4 = 1$ or 7

(67 and 469 are eliminated by the fact that each child ate at least one chocolate, which implies that $c \geq n$.)

Thus, n is either 5 or 11. But 5 is impossible since six children are mentioned by name, namely Evelyn, Charles, Kitty, Clare, Edna, and David. Therefore, 11 children ate 67 chocolates, and 67 can be partitioned into 11 different numbers in exactly one way:

$$1, 2, 3, 4, 5, 6, 7, 8, 9, 10, 12.$$

Therefore, 11 children in all ate the chocolates. Edna ate 12.

(168) *Problem.* In cleaning out the top drawer of her dresser, Mrs. Adams found two wrist watches she and her husband had long ago discarded as impossibly inaccurate timepieces. She decided to give them one last trial. Therefore she wound them carefully and, after setting them accurately, started both watches going at the same time. Then she slipped her old watch on one wrist and her husband's on the other wrist so that she might keep checking them at odd intervals during the day. An hour later she noticed that while her watch had gained one minute her husband's had lost two minutes.

Next morning when she looked at the watches again, it was seven o'clock by her old watch and six o'clock by her husband's.

What time was it when she started the watches running?

Solution. It is intended that we assume the watches continued to run at a constant rate. Then, every hour Mrs. Adams' watch gains 3 minutes on her husband's, which means it would take 20 hours for the 2 watches to be exactly 1 hour different. During this 20 hours Mrs. Adams' watch will have gained 20 minutes (and her husband's watch will have lost 40 minutes), so that actual time in the morning is 6:40 am. Twenty hours earlier it was 10:40 am, which is the time she started both watches going.

(169) *Problem.* Imagine that you have three boxes, one containing two black marbles, one containing two white marbles, and the third, one black marble and one white marble. The boxes are labelled for their contents BB, WW, and BW, but someone has switched the labels so that every box is now incorrectly labelled. You are allowed to draw one marble at a time out of any box, without looking inside, and by this process of sampling you are to determine the contents of all three boxes.

What is the smallest number of draws needed to do this?

Solution. You can learn the contents of all three boxes by drawing just one marble. The key to the situation is your knowledge that the labels on all three of the boxes are incorrect. You must draw a marble from the box labelled BW. Assume that the marble drawn is black. You know that the other marble in this box must be black also, otherwise the label would be correct. Since you have now identified the box containing the two black marbles, you can at once tell the contents of the box labelled WW: you know it cannot contain two white marbles, because its label has to be wrong; it cannot contain two black marbles, for you have identified that box; therefore it must be contain one black and one white marble. The third box must then be the one holding two white marbles. A similar analysis can be applied if the marble originally drawn from the BW box was white instead of black.

(170) *Problem.*   Twelve squares are laid out in a circular pattern [as on the circumference of a circle]. Four different coloured chips, red, yellow, green, and blue are placed on four consecutive squares. A chip may be moved in either a clockwise or counterclockwise direction over four other squares to a fifth square, provided the fifth square is not occupied by a chip. After a certain number of moves the same four squares will again be occupied by chips. How many possible rearrangements of the four chips can occur as a result of this process?

Solution.   Let us first number the squares from 1 to 12, starting, say, from the red chip and continuing through the chips and around the circle. Now let us rearrange the squares by putting them in an order in which it is possible to move a chip from one square to the next. That is, after square 1 we place square 6 (since by the conditions of the problem it is possible for a chip to move from square 1 to square 6, and vice versa); then after square 6, we place square 11 (since a chip may move from square 6 to 11), and after square 11, we place square 4, and so on. After this rearrangement we have the following order of squares:

$$
\begin{array}{ccccc}
R & & B & & Y \\
1 \longleftrightarrow 6 \longleftrightarrow 11 \longleftrightarrow 4 \longleftrightarrow & 9 \longleftrightarrow 2 \\
\updownarrow & & & & \updownarrow \\
8 \longleftrightarrow 3 \longleftrightarrow 10 \longleftrightarrow 5 \longleftrightarrow 12 \longleftrightarrow 7 \\
G & & & &
\end{array}
$$

The chips (red, yellow, green, blue, designated R, Y, G, B) are shown adjacent to their original positions: R on 1, Y on 2, G on 3, and B on 4. The rule by which the chips may now move is simple: A chip may move one square in either direction (in the above diagram), provided that the new square is unoccupied. Thus, the only way in which a chip can change places with another chip is for it to move around the above rectangle in either direction, but now a chip can neither jump over nor occupy the position of another chip. Thus, if chip R is moved to occupy square 4, then B must occupy square 2, Y must occupy square 3, and G must move to square 1. Similar arguments show there is only one arrangement if R moves to occupy square 3, or if R moves to occupy square 2. Thus, there is a total of 4 possible rearrangements of the 4 coloured chips on the squares numbered 1 to 4:

$$RYGB, BRYG, GBRY, YGBR.$$

(These are sometimes called *cyclic permutations* of $RYGB$.)

Chapter 6

Number Theory

(171) *Problem.* One and only one of the following numbers is a perfect square. Which is it? Why? Do <u>not</u> compute the square roots. In fact, do <u>not</u> use calculators or computers at all.

$$3{,}669{,}517{,}136{,}205{,}224$$
$$1{,}898{,}732{,}825{,}398{,}318$$
$$4{,}715{,}006{,}864{,}295{,}101$$
$$5{,}901{,}643{,}220{,}186{,}100$$
$$7{,}538{,}062{,}944{,}751{,}882$$
$$2{,}512{,}339{,}789{,}576{,}516$$

Solution. Let us suppose that the integer a is the square root of the number we wish to identify. Then a can be written as $10b + c$, where b is some natural number and c is a digit between 0 and 9, inclusive. Then

$$a^2 = (10b + c)^2 = 100b^2 + 20bc + c^2 = 10(10b^2 + 2bc) + c^2.$$

Thus, the units digit of a^2 is affected only by the units digit of c^2. Since the units digit of c^2 can only be one of 0, 1, 4, 5, 6, or 9, we have eliminated two of the possibilities.

Now let us write a in the form $9d + e$, where d is a natural number and e is a digit between 0 and 8, inclusive. Then

$$a^2 = (9d + e)^2 = 81d^2 + 18de + e^2 = 9(9d^2 + 2de) + e^2.$$

Thus, on division by 9 the number we are seeking must leave a remainder of e^2 (actually e^2 reduced by some multiple of 9 in order to get a remainder in the proper range from 0 to 8, inclusive). Now e^2, on division by 9, leaves a remainder which is one of 0, 1, 4, or 7. Consequently, a^2 must leave a remainder of 0, 1, 4, or 7 on division by 9. But we know that the remainder on dividing a number by 9 is the same as the remainder on dividing the sum of its digits by 9. Of the four remaining candidates, the sum of the digits has a remainder in the set $\{0, 1, 4, 7\}$ only for 2,512,339,789,576,516. (It is, in fact, the square of 50,123,246.)

(172) *Problem.*   Bob's stamp collection consists of three books. Two-tenths of his stamps are in the first book, several sevenths in the second book, and there are 303 stamps in the third book.

Is there enough information to determine how many stamps Bob has?

· If so, how many has he? If not, why not? (That is, explain why not.)

Solution.   Let x be the number of stamps in Bob's collection, and let y be the number of sevenths in the second book. Clearly, $y \geq 3$. Then

$$x = \frac{2x}{10} + \frac{yx}{7} + 303 = \frac{x}{5} + \frac{yx}{7} + 303.$$

Clearing the fractions yields:

$$35x = 7x + 5yx + 35(303),$$
$$28x - 5yx = 35(303),$$
$$x(28 - 5y) = 3 \cdot 5 \cdot 7 \cdot 101,$$

where the factors on the right side of the last line are all prime numbers. Clearly, $28 - 5y \geq 0$, which implies that $y \leq 5$.

The only acceptable value of y which yields a value of $28 - 5y$ in the set $\{3, 5, 7, 101\}$ is $y = 5$.

Then $28 - 5y = 3$, and $x = 5 \cdot 7 \cdot 101 = 3535$. Hence, Bob has 3535 stamps in his collection.

It is interesting to note that we never actually needed $y \geq 3$. We only used $y \geq 0$.

(173) *Problem.*   The expression of a positive integer n in base b is

$$n = (1254)_b.$$

It is known that the expression of the integer $2n$ in the same base is

$$2n = (2541)_b.$$

Determine the values of b and n in base 10.

Solution. The number $(1254)_b$ is $b^3 + 2b^2 + 5b + 4$ and the number $(2541)_b$ is $2b^3 + 5b^2 + 4b + 1$. Consequently,

$$n = b^3 + 2b^2 + 5b + 4, \tag{1}$$

$$2n = 2b^3 + 5b^2 + 4b + 1. \tag{2}$$

Multiplying (1) by 2 and comparing with (2) yields

$$5b^2 + 4b + 1 = 4b^2 + 10b + 8,$$

$$b^2 - 6b - 7 = 0,$$

$$(b - 7)(b + 1) = 0,$$

which means that $b = 7$ or $b = -1$. Since a base is always positive, we must have $b = 7$. Therefore,

$$n = (1254)_7 = 7^3 + 2 \cdot 7^2 + 5 \cdot 7 + 4$$

$$= 343 + 98 + 35 + 4$$

$$= 480.$$

This solution is perhaps harder than necessary since $1254 + 1254$ must equal 2541, so that 4 and 4 must have a left over of 1, indicating that is must be base 7. Checking in base 7, it works.

(174) *Problem.* A number (say, of less than 25 digits) begins with the digits 15. When it is multiplied by 5, the result is merely to shift those digits to the right end.

What is the number?

Solution. Let us suppose the unknown part of the number N is the number A having a digits. Then, by considering the positional notation in base 10, we have

$$N = 15 \times 10^a + A.$$

But we also have

$$5N = A \times 100 + 15.$$

By eliminating A, we get

$$95N = 15 \times 10^{a+2} - 15$$

$$\text{or} \qquad 19N = 3 \times 10^{a+2} - 3 = \underbrace{2999 \cdots 997}_{a+1 \text{ 9's}}.$$

In carrying out this long division, we bring down 9 continually, unless in bringing down 7 we would get even division. Solving this, we get

$$N = 157894736842105263,$$

a number with 18 digits. The next possibility is simply the above number repeated twice (that is, 36 digits), which has too many digits!

(175) *Problem.* How old is the captain, how many children has he, and how long is his boat, given the product 32118 of the three desired integers? The length of his boat is given in feet (several feet), the captain has both sons and daughters, he has more years than children, but he is not yet one hundred years old. (Give reasons for your answer.)

Solution. The number 32118 can be factored into primes as

$$32118 = 2 \cdot 3 \cdot 53 \cdot 101.$$

If we let x be the captain's age in years, y be the number of his children, and z the length of his boat in feet, we see immediately that

$$y < x \le 99.$$

If we further notice that $y \ge 4$ (since the dictionary meaning of "several" is "at least 4"), then we have

$$4 \le y < x \le 99.$$

The only possible choices for x and y are $x = 53$ and $y = 6$. This leaves $z = 101$. Thus, the captain is 53 years old, has 6 children, and a boat which is 101 feet in length.

(176) *Problem.* If x is an integer, and if $2x + 1$ and $3x + 1$ are both perfect squares, show that x is a multiple of 40.

Solution. Let $a^2 = 2x + 1$ and $b^2 = 3x + 1$ be our given perfect squares. By rearranging and factoring, we obtain

$$2x = a^2 - 1 = (a - 1)(a + 1), \tag{1}$$
$$3x = b^2 - 1 = (b - 1)(b + 1). \tag{2}$$

Notice that the numbers $a-1$ and $a+1$ are either both even or both odd, since they differ by 2; the same is true of $b-1$ and $b+1$. From equation (1) we see that both $a - 1$ and $a + 1$ must be even. Thus, the right side of equation (1) is a multiple of 4. Consequently, x is even.

From equation (2) and the fact that x is even, we observe that both $b - 1$ and $b + 1$ are even. Not only that, but they are consecutive even numbers. This means that one of them is a multiple of 4.

Thus, the right side of equation (2) is a multiple of 8. Therefore, x must be a multiple of 8.

To finish the solution, we need only show that x is a multiple of 5. We will proceed in steps:

1. Any perfect square n^2 gives a remainder of 0, 1, or 4 when divided by 5.

Proof: Let $n = 5k + c$, where $c \in \{0, 1, 2, 3, 4\}$ (that is, c is the remainder when n is divided by 5). Then

$$n^2 = (5k + c)^2 = 25k^2 + 10kc + c^2 = 5(5k^2 + 2kc) + c^2,$$

and $c^2 \in \{0, 1, 4, 9, 16\}$. Since the numbers 9 and 16 leave remainders of 4 and 1, respectively, when divided by 5, the proof is complete.

2. Let $x = 5k + d$, where $d \in \{0, 1, 2, 3, 4\}$ (that is, d is the remainder when x is divided by 5). We must show that $d = 0$ is the only possibility for our given value of x.

(a) $2d + 1$ has a remainder of 0, 1, or 4 when divided by 5.
Proof: First we have $a^2 = 2x + 1 = 2(5k + d) + 1 = 10k + 2d + 1$. From Step 1, we see that $2d + 1$ has a remainder of 0, 1, or 4 when divided by 5.

(b) $3d + 1$ has a remainder of 0, 1, or 4 when divided by 5.
Proof: First we have $b^2 = 3x + 1 = 3(5k + d) + 1 = 15k + 3d + 1$. From Step 1, we see that $3d + 1$ has a remainder of 0, 1, or 4 when divided by 5.

(c) $d = 0$.
Proof: From Step 2(a), we must have $d \in \{0, 2, 4\}$; from Step 2(b), we must have $d \in \{0, 1, 3\}$. The only common choice is $d = 0$.

Therefore, x is a multiple of 5 and a multiple of 8, which proves that x is a multiple of 40.

(177) *Problem.* How many zeroes are there at the end of the number

$$1000! = 1 \cdot 2 \cdot 3 \cdots \cdot 1000?$$

If we were dealing with base eight arithmetic what would the answer be? What about base twelve?

Solution. A new zero will occur in the factorial computation every time a "5" and a "2" appear together. Since 2's appear more frequently than 5's (that is, even numbers are more numerous than multiples of 5), we need only concern ourselves with the number of 5's that appear in the factorizations of the numbers from 1 to 1000.

Now, every multiple of 5 contributes at least one such 5. This gives us 200 fives to start with. However, this only accounts for one of the 5's in the multiples of 25, but each multiple of 25 contributes at least two 5's.

Thus, we have an additional 40 fives from the multiples of 25. By considering multiples of 125 and 625, we get eight more fives and one more five, respectively. This gives a grand total of 249 fives.

Therefore, 1000! ends in 249 zeroes. If one uses $\lfloor x \rfloor$ to denote the greatest integer not exceeding x, we see that the number k of zeroes that $n!$ ends in is given by

$$k = \left\lfloor \frac{n}{5} \right\rfloor + \left\lfloor \frac{n}{5^2} \right\rfloor + \left\lfloor \frac{n}{5^3} \right\rfloor + \cdots .$$

If $n = 1000$, this produces

$$k = \left\lfloor \frac{1000}{5} \right\rfloor + \left\lfloor \frac{1000}{25} \right\rfloor + \left\lfloor \frac{1000}{125} \right\rfloor + \left\lfloor \frac{1000}{625} \right\rfloor + \cdots$$

$$= 200 + 40 + 8 + 1 + 0 + \cdots = 249.$$

Now assume we are dealing with base eight. That is, how many zeroes does $(1000)_8!$ end in? By arguing as in base ten above, this will depend entirely upon the number of 2's appearing in the factorizations of the numbers from 1 to $(1000)_8$. Every grouping of three 2's will produce a zero. To proceed further, let us convert $(1000)_8$ to base ten.

$$(1000)_8 = 1 \times 8^3 + 0 \times 8^2 + 0 \times 8^1 + 0 \times 8^0 = 512.$$

If k is the number of 2's appearing in 512!, then as above we have

$$k = \left\lfloor \frac{512}{2} \right\rfloor + \left\lfloor \frac{512}{4} \right\rfloor + \left\lfloor \frac{512}{8} \right\rfloor + \left\lfloor \frac{512}{16} \right\rfloor + \cdots + \left\lfloor \frac{512}{512} \right\rfloor + \cdots$$

$$= 256 + 128 + 64 + 32 + 16 + 8 + 4 + 2 + 1 + 0 + \cdots = 511.$$

Since three 2's are required to produce each zero, we then have $\lfloor \frac{511}{3} \rfloor = 170$ zeroes at the end of $(1000)_8!$ in base eight.

Note: If you had interpreted 1000! to already be in base ten notation, then the above computation would yield

$$k = \left\lfloor \frac{1000}{2} \right\rfloor + \left\lfloor \frac{1000}{4} \right\rfloor + \left\lfloor \frac{1000}{8} \right\rfloor + \left\lfloor \frac{1000}{16} \right\rfloor + \cdots + \left\lfloor \frac{1000}{512} \right\rfloor + \cdots$$

$$= 500 + 250 + 125 + 62 + 31 + 15 + 7 + 3 + 1 + 0 + \cdots = 994,$$

and thus, $\left\lfloor \frac{994}{3} \right\rfloor = 331$ zeroes.

With base twelve we have a slightly more difficult problem, since we need two 2's and one 3 to produce each zero. We would like to know which occurs less rapidly. Instead, we will determine the number of pairs of 2's and the number of 3's independently; the smaller of these two numbers will be the number of zeroes. First of all we note that

$$(1000)_{12} = 1 \times 12^3 + 0 \times 12^2 + 0 \times 12^1 + 0 \times 12^0 = 1728.$$

Let k_2 and k_3 be the number of 2's and 3's, respectively, appearing in 1728!. Then

$$k_2 = \left\lfloor \frac{1728}{2} \right\rfloor + \left\lfloor \frac{1728}{4} \right\rfloor + \left\lfloor \frac{1728}{8} \right\rfloor + \left\lfloor \frac{1728}{16} \right\rfloor + \cdots + \left\lfloor \frac{1728}{1024} \right\rfloor + \cdots$$

$$= 864 + 432 + 216 + 108 + 54 + 27 + 13 + 6 + 3 + 1 + 0 + \cdots$$

$$= 1724,$$

$$k_3 = \left\lfloor \frac{1728}{3} \right\rfloor + \left\lfloor \frac{1728}{9} \right\rfloor + \left\lfloor \frac{1728}{27} \right\rfloor + \left\lfloor \frac{1728}{81} \right\rfloor + \cdots + \left\lfloor \frac{1728}{729} \right\rfloor + \cdots$$

$$= 576 + 192 + 64 + 21 + 7 + 2 + 0 + \cdots = 862.$$

Therefore, the number of pairs of 2's is $\lfloor 1724/2 \rfloor = 862$. (Interestingly enough the number of pairs of 2's and the number of 3's is the same.) Thus, there are 862 zeroes at the end of $(1000)_{12}!$ in base twelve.

Note: If you had interpreted 1000 to already be in base 10 notation, then the above computation would yield

$$k_2 = \left\lfloor \frac{1000}{2} \right\rfloor + \left\lfloor \frac{1000}{4} \right\rfloor + \left\lfloor \frac{1000}{8} \right\rfloor + \left\lfloor \frac{1000}{16} \right\rfloor + \cdots + \left\lfloor \frac{1000}{512} \right\rfloor + \cdots$$

$$= 500 + 250 + 125 + 62 + 31 + 15 + 7 + 3 + 1 + 0 + \cdots = 994,$$

$$k_3 = \left\lfloor \frac{1000}{3} \right\rfloor + \left\lfloor \frac{1000}{9} \right\rfloor + \left\lfloor \frac{1000}{27} \right\rfloor + \left\lfloor \frac{1000}{81} \right\rfloor + \cdots + \left\lfloor \frac{1000}{729} \right\rfloor + \cdots$$

$$= 333 + 111 + 37 + 12 + 4 + 1 + 0 + \cdots = 498.$$

Therefore, the number of pairs of 2's is $\lfloor 994/2 \rfloor = 497$, and thus, 497 zeroes.

(178) *Problem.*   Show that the expressions
$$2x + 3y \qquad \text{and} \qquad 9x + 5y$$
are divisible by 17 for the same set of integer values of x and y.
Solution.   Note first that
$$3(9x + 5y) - 5(2x + 3y) = 17x.$$
By rewriting this as
$$3(9x + 5y) = 5(2x + 3y) + 17x,$$
we see that whenever 17 divides $2x + 3y$, it must divide $9x + 5y$ also. And by rewriting it as
$$5(2x + 3y) = 3(9x + 5y) - 17x,$$
we see that whenever $9x + 5y$, it must divide $2x + 3y$ also. The problem is finished!

Note: This problem can be generalized to:
"Show that if $|ad - bc| = p$, a prime number, then $ax + by$ and $cx + dy$ are divisible by p for the same set of integer values of x and y."

(179) *Problem.*   Among grandfather's papers a bill was found:

72 turkeys

$_76.9_

The first and last digits of the number that obviously represented the total price of those fowls are replaced here by blanks, for they have faded and are now illegible. What are the two faded digits, and what was the price of one turkey?
Solution.   We will make the assumption that each turkey costs the same amount. Then 72 must evenly divide the amount of money on the bill. Since $72 = 9 \times 8$, both 9 and 8 evenly divide the amount of money. Let a be the first (missing) digit, and let b be the last (missing) digit.

If we convert the money on the bill to pennies, we have $a679b$ pennies, which is $a6 \times 1000 + 67b$. Since 8 divides 1000 evenly, it must also evenly divide $79b$. This means that $b = 2$. If 9 evenly divides any number, then it also evenly divides the sum of the digits of that number. Thus, 9 divides $a + 6 + 7 + 9 + 2 = a + 24$. Hence, $a = 3$.

The two faded digits are 3 and 2 and the cost of one turkey is
$$\$367.92/72 = \$5.11.$$

(180) *Problem.* Let n be an integer. If the tens digit of n^2 is 7, what is the units digit of n^2?

Give reasons.

Solution. Let a and b be the final two digits of n. Then we wish to know the last digit of n^2, which is the same as the last digit of b^2. The last two digits of n^2 will receive contributions only from a and b. In fact, the last two digits of n^2 will be the same as the last two digits of $(10a + b)^2 = 100a^2 + 20ab + b^2$, and hence, leaves a remainder of b^2 upon division by 20. Since we know the tens digit is a 7, we require a remainder from 10 to 19 when divided by 20. The only possible final digit is 6. (This can be checked by considering the squares of $0, 1, 2, 3, \ldots, 9$ giving 16 as the only remainder satisfying this requirement.)

(181) *Problem.* In the Canadian *Lotto 6/49* a ticket consists of 6 distinct integers chosen from 1 to 49 (inclusive). Prizes are awarded for having 3, 4, 5, or 6 numbers in common with a designated "winning" ticket.

Let us generalize this a bit further: a t-prize will be awarded for any ticket having t or more numbers in common with the "winning" ticket. Define by $f(t)$ the smallest number of tickets required to be certain of winning a t-prize.

Note that $f(1) = 8$ since 7 tickets can cover at most 42 numbers. Also note that $f(6) = \binom{49}{6}$, which is 13,983,816, since every possible combination of 6 numbers is a potential winner.

Show that $f(2) \leq 19$; that is, one never needs to buy more than 19 tickets in order to guarantee at least two numbers on one ticket in common with the "winning" ticket.

Solution. First of all notice that by considering the following tickets:

$$1\ 2\ 3\ 4\ 5\ \ 6,$$
$$1\ 2\ 7\ 8\ 9\ 10,$$
$$3\ 4\ 7\ 8\ 9\ 10,$$
$$5\ 6\ 7\ 8\ 9\ 10,$$

we see that one needs at most 4 tickets in order to guarantee that at least one of the tickets has any two possible numbers from 1 to 10 (inclusive). A similar argument would work on any set of 10 numbers. By considering the 3 tickets:

$$1\,2\,3\,4\,5\,6,$$
$$1\,2\,3\,7\,8\,9,$$
$$4\,5\,6\,7\,8\,9,$$

we see that if the upper bound is 9, it suffices to have 3 tickets. Now divide the integers from 1 to 49 into 5 groups, namely from 1 to 10, from 11 to 20, from 21 to 30, from 31 to 40, and from 41 to 49 (all inclusive). At least one of these groups has at least two or more numbers in common with the "winning" ticket. Hence,

$$f(2) \leq 4 \cdot 4 + 3 = 19.$$

(182) *Problem.*   Professor Adams wrote on the blackboard a polynomial, $f(x)$, with integer coefficients and said, "Today is my son's birthday, and when we substitute x equal to his age A, then $f(A) = A$. You will also note that $f(0) = p$, and that p is a prime number greater than A."

How old is Professor Adams' son?

Solution.   Let $f(x) = a_0 x^n + a_1 x^{n-1} + \cdots + a_{n-1} x + a_n$. Since $f(0) = p$, we have $a_n = p$. Also, $f(A) = A$. Therefore,

$$a_0 A^n + a_1 A^{n-1} + \cdots + a_{n-1} A + p = A$$
$$A(a_0 A^{n-1} + a_1 A^{n-2} + \cdots + a_{n-1}) = A - p.$$

Therefore, A divides $A - p$ evenly, which means that A divides p evenly. Since p is a prime greater than A, we must have $A = 1$. Hence, Professor's Adams' son is 1 year old.

(183) *Problem.*   The factorial of a positive integer n is defined as

$$n! = n \cdot (n-1) \cdot (n-2) \cdot \cdots \cdot 3 \cdot 2 \cdot 1$$

(that is, the product of the first n positive integers). It is sometimes defined inductively as

$$1! = 1,$$
$$n! = n \cdot (n-1)!, \qquad \text{for } n \geq 2.$$

Show that $n! + 1$ and $(n+1)! + 1$ have no common factor (other than 1).

Solution.   Let $n! + 1$ and $(n+1)! + 1$ have a common factor $d > 0$. We must show that $d = 1$. Let $a = n! + 1$ and $b = (n+1)! + 1$. Then

$$b = (n+1)! + 1 = (n+1) \cdot n! + 1 = (n+1)(a-1) + 1$$
$$= (n+1)a - n - 1 + 1 = (n+1)a - n.$$

Therefore, $n = (n+1)a - b$. Since n is a factor of both a and b, it must clearly be a factor of n. But

$$a = n! + 1 = n \cdot (n-1)! + 1$$

$$\text{or} \quad 1 = a - n \cdot (n-1)!$$

Since d is a factor of both a and n, it must be a factor of 1. Thus, $d = 1$. Therefore, $n! + 1$ and $(n+1)! + 1$ have no common factor (other than 1).

(184) *Problem.* We know a young chick in Dundee,

Whose age has its last digit "three".

The square of the first

Is her whole age reversed.

So what must the lady's age be?

Solution. Let the age of the "young chick" be $10a + b$, where a and b are both integers between 0 and 9. Then clearly $b = 3$. Also, we are told that

$$a^2 = 10b + a = a + 30.$$

Then $a^2 - a - 30 = 0$, from which we see that $(a-6)(a+5) = 0$. Thus, $a = 6$ or -5. Therefore, a must be 6 and the age we seek is 63.

(185) *Problem.* Rectangles are constructed with 1 cm square tiles. If the measure (in cm^2) of the area of the rectangle is equal to the measure (in cm) of the perimeter of the rectangle, the rectangle is called a "balanced rectangle". For example, a 4×4 rectangle is balanced. Are there more balanced rectangles? If yes, find the dimensions of all other balanced rectangles. If no, explain why not.

Solution. Let a and b be the dimensions of the balanced rectangle. Then we are clearly looking for a solution (a, b) in positive integers to the equation

$$ab = 2(a + b).$$

Since a and b are integers, we can draw the following conclusions:

1. ab is even.
2. either a is even or b is even (or both).

Since a and b play symmetric roles, let us assume that a is even, say $a = 2k$. Then we have

$$2kb = 2(2k + b)$$

$$\text{or} \quad kb = 2k + b$$

$$\text{that is,} \quad (k-1)b = 2k.$$

Since $k - 1$ is a factor of the left side, it must be a factor of the right side. But $k - 1$ and k are two consecutive integers, and have no common factor. This implies that $k - 1$ is a divisor of 2. That is, $k - 1 = 1$ or $k - 1 = 2$.

(i) $k - 1 = 1$. This implies that $k = 2$ and $a = b = 4$.

(ii) $k - 1 = 2$. This implies that $k = 3$, $b = 3$, and $a = 6$.

Thus, the only balanced rectangles have dimensions 4×4 and 6×3.

(186) *Problem.*   The enrollment in a certain undergraduate mathematics course consists of sophomores, juniors and seniors. If each sophomore contributes $1.25, each junior $0.90, and each senior $0.50, the class will have a fund of $25.

If there are 26 students in the class, how many of each kind are there?

Solution.   Let x, y, z represent the numbers of sophomores, juniors and seniors, respectively, in the above mathematics class. If we express all monetary amounts in the form of cents, we can generate the following equation:

$$125x + 90y + 50z = 2500.$$

In addition, since there are 26 students in the class, we also have the equation:

$$x + y + z = 26.$$

By eliminating the variable z in these two equations, we obtain a new equation:

$$15x + 8y = 240,$$

from which it is possible to deduce that 15 divides evenly into y and that 8 divides evenly into x. Thus, x is in the set $\{0, 8, 16, 24\}$ and y is in the set $\{0, 15\}$. By starting with the original equations above and eliminating the variable y, we obtain the new equation:

$$7x - 8z = 32, \quad \text{or} \quad 7x = 8(4 + z),$$

from which we can see that 7 must divide evenly into $4 + z$. Thus, z belongs to the set $\{3, 10, 17, 24\}$. The only possible combinations of values which satisfy $x + y + z = 26$ are:

(i) $x = 8$, $y = 15$, $z = 3$, and

(ii) $x = 16$, $y = 0$, $z = 10$.

However, when one considers that the statement of the problem says that the class includes students in each category, we see that only case (i) is allowed.

(187) *Problem.* We have eight boxes – one red, the others blue and yellow – each containing a different number of balls: 11, 14, 19, 23, 29, 32, 41, and 46. The total of all balls in yellow boxes is twice the total of all balls in blue boxes. How many balls are in the red box?

Solution. The first thing to notice is that there are 215 balls altogether. Let Y be the number of balls in the yellow boxes, B the number in the blue boxes and R the number in the red boxes. Then we certainly have $Y = 2B$. Thus,

$$R + Y + B = R + 3B = 215,$$

$$\text{or} \qquad R = 215 - 3B = 3(71 - B) + 2.$$

This means that R must leave a remainder of 2 when divided by 3. Unfortunately, this only rules out the values 19 and 46. There remains still six possibilities for R, each of which fully determines the values for Y and B:

R	Y	B	Non-red numbers
11	136	68	14, 19, 23, 29, 32, 41, 46
14	134	67	11, 19, 23, 29, 32, 41, 46
23	128	64	11, 14, 19, 29, 32, 41, 46
29	124	62	11, 14, 19, 23, 32, 41, 46
32	122	61	11, 14, 19, 23, 29, 41, 46
41	116	58	11, 14, 19, 23, 29, 32, 46

We must now find some combination of the number of balls (other than the number selected as a candidate for R) which adds up to the corresponding value of B. By a process of exhaustion, the only choice that works is $R = 29$.

Addendum: Note that the blue boxes contain 11, 19 and 32 for a total $B = 62$.

(188) *Problem.* An M-athlon is a competition in which there are M athletic events. Such a competition was held in which only A, B, and C participated. In each event p_1 points were awarded for first place, p_2 for second, and p_3 for third, where $p_1 > p_2 > p_3 > 0$

and p_1, p_2, and p_3 are all integers (that is, whole numbers). The final score for A was 22, for B was 9, and for C was 9. B won the 100 metres. What is the value of M and who was second in the high-jump?

Solution. For each of the M events there is a total of $p_1 + p_2 + p_3$ points awarded. Thus, for the entire set of M events there is a total of $M(p_1 + p_2 + p_3)$ points, which must be the same as the total of the points received, which is $9 + 9 + 22 = 40$. Therefore,

$$M(p_1 + p_2 + p_3) = 40. \tag{1}$$

Since we are dealing with integer arithmetic, we may conclude that M evenly divides into 40. Since there are two events which have been named in the statement of the problem, M must be at least 2. Let us now consider B. In addition to winning the 100 metres, the worst he could have done was to have finished last in the remaining $M - 1$ events. Thus,

$$p_1 + (M - 1)p_3 \leq 9, \tag{2}$$

whence we conclude that $p_1 < 9$.

Also since $p_1 > p_2 > p_3 > 0$, we see that $p_1 + p_2 + p_3 \geq 6$. Thus, from (1) above we may conclude that $M \leq 6$ and divides evenly into 40. This implies that M is one of 2, 4, or 5. However, $p_1 < 9$, which means that there must be at least 3 events in order for A to get 22 points. Thus, M is 4 or 5.

Now for M at least 4, equation (2) above yields $p_1 \leq 6$. If we assume for the moment that $M = 4$, the fact that A received 22 points implies that $p_2 \geq 4$ and $p_1 = 6$, which will now contradict equation (1). Thus, M must be 5. Equation (2) then forces p_1 to be 5 (otherwise A's total of 22 cannot be reached) and p_3 to be 1. Equation (1) then yields $p_2 = 2$. Furthermore, to achieve a total of 22 points, A must have finished first in 4 events and second in the other.

It can then be seen that C finished second in every event except the 100 metres, that is, C finished second in the high jump!

(189) *Problem.* "This might interest you, professor," said John. "My age and the ages of each of my three distant cousins are all prime numbers, and the sum of our ages is 50."

"In that case," said the professor, who knew John's age, "I can tell you the ages of your three cousins."

You do not share the professor's advantage of knowing John's age to start with, but nevertheless, can you tell the ages of his cousins? (Note that the number 1 is not considered to be prime.)

Solution. The first step in solving this problem is to list all the primes less than 50. They are

$$2, 3, 5, 7, 11, 13, 17, 19, 23, 29, 31, 37, 41, 43, 47.$$

The next step is to notice that the ages need not be distinct! If we let John's age be x, then the statement made by the professor indicates that $50 - x$ can be written uniquely as a sum of three primes.

Let us now consider the different possible values of x to discover if we can find more than one set of three primes which sum to $50 - x$.

x	$50 - x$	Sets of primes summing to $50 - x$
2	48	$\{2, 3, 43\}$, $\{2, 5, 41\}$, and more
3	47	$\{2, 2, 43\}$, $\{3, 3, 41\}$, and more
5	45	$\{2, 2, 41\}$, $\{3, 5, 37\}$, and more
7	43	$\{3, 3, 37\}$, $\{3, 11, 29\}$, and more
11	39	$\{3, 5, 31\}$, $\{3, 7, 29\}$, and more
13	37	$\{3, 3, 31\}$, $\{3, 5, 29\}$, and more
17	33	$\{2, 2, 29\}$, $\{3, 7, 23\}$, and more
19	31	$\{3, 5, 23\}$, $\{3, 11, 17\}$, and more
23	27	$\{2, 2, 23\}$, $\{3, 5, 19\}$, and more
29	21	$\{2, 2, 17\}$, $\{3, 5, 13\}$, and more
31	19	$\{3, 3, 13\}$, $\{3, 5, 11\}$, and more
37	13	$\{3, 3, 7\}$ and $\{3, 5, 5\}$
41	9	$\{2, 2, 5\}$ and $\{3, 3, 3\}$
43	7	$\{2, 2, 3\}$, but no others!
47	3	no such sets!

From the above table we can conclude that John is 43 and his cousins are ages 2, 2, and 3.

(190) *Problem.* At harvest time, the orchards of Mr. MacIntosh, Mr. Jonathan, and Mr. Delicious had yielded 314,827 apples, 1,199,533 apples, and 683,786 apples, respectively. While lunching with Jonathan the following Sunday, MacIntosh mentioned the number of apples he would have left over if he divided his harvest equally among all the apple dealers.

"Why don't you sell those extra apples to me," suggested Jonathan, "and then I'll be able to divide my apples equally among all the apple dealers"

"Sorry," said MacIntosh, "but Mr. Delicious made the same suggestion for the same reason, and I've already accepted his offer."

How many apple dealers are there?

Solution. Let n be the number of dealers, and let m, j and d be the number of apples each of MacIntosh, Jonathan and Delicious, respectively, would give to each dealer if he divided his apples equally among the dealers. Then they would have left

$$314{,}827 - nm,$$
$$1{,}199{,}533 - nj,$$
$$\text{and} \quad 683{,}786 - nd,$$

respectively, where each of these remainders is an integer between 0 and $n - 1$. This implies that

$$(1{,}199{,}533 - nj) + (314{,}827 - nm) \text{ is a multiple of } n,$$
$$\text{as is } (683{,}786 - nd) + (314{,}827 - nm),$$

from which we can conclude that both 1,514,360 and 998,613 are multiples of n. Using the Euclidean algorithm we find the greatest common divisor of these numbers is 131, which is a prime. Since n must divide the greatest common divisor, n must be 131 or 1. Clearly 1 is not an acceptable answer, so that the answer is 131 dealers.

(191) *Problem.* In the following, each of the letters A, B, C and D represents a different digit in base 10.

$$(AB) \cdot (CB) = DDD.$$

(Here AB means $10A + B$, and not the product of A and B. But the dot between (AB) and (CB) does indicate multiplication.)

What are the three numbers AB, CB and DDD where $A > C$?

Solution. The factors of the right hand side of the given equation are D and 111. Since 111 factors into 3 and 37, the integers D, 3 and 37 must all divide the left hand side. Since 37 is prime, one of the two values, say AB must be either 37 or 74. If $AB = 74$, then $B = 4$, whence $D = 6$, which fails to work. Thus, AB must be 37 and $D = 9$. The solution, therefore, is

$$AB = 37$$
$$CB = 27$$
$$DDD = 999.$$

(192) *Problem.* A professor was interviewing a prospective graduate assistant for his class in number theory.

"I'm going to test you with a little puzzle," the professor said. "In my neighbourhood are four families with children. I have the largest number of children; the Browns have a smaller number; the Greens have a still smaller number, and the Blacks have the smallest number of all. There are fewer than fifteen children, and the product of these four numbers is my house number. How many children are in each family?"

"I have forgotten your house number," replied the student.

"That's all right," said the professor. "Even if you had remembered it, you still wouldn't have enough information to solve this puzzle, so I'll give you a hint. The number of children in the Green family is equal to the number of times you have dated my daughter."

Believe it or not, the student was able to solve the puzzle with only this scant information. How many children were in each family?

Solution. Let a, b, c, and d be the number of children in the professor's, the Brown's, the Green's and the Black's families, respectively. Clearly, we can state

$$a > b > c > d \quad \text{and} \quad a + b + c + d < 15.$$

Let us build a table of possibilities and include in the table the value of the product.

a	b	c	d	Product (house number)
4	3	2	1	24
5	3	2	1	30
6	3	2	1	36
7	3	2	1	42
8	3	2	1	48
5	4	2	1	40
6	4	2	1	48
7	4	2	1	56
5	4	3	1	60
6	4	3	1	72
5	4	3	2	100
6	5	2	1	60

If the house number was neither 48 nor 60, then this information would have been enough to determine the answer, but we are told that this information is insufficient. Thus, the house number must be either 48 or 60. Let us consider those cases again:

a	b	c	d	Product (house number)
8	3	2	1	48
6	4	2	1	48
5	4	3	1	60
6	5	2	1	60

Since knowing the value of c (the number of times he dated the professor's daughter) allows the student to solve the problem, the value of c must be 3, and the number of children are 5, 4, 3, and 1 for the professor, the Browns, the Greens, and the Blacks, respectively.

(193) *Problem.*   "Funny," says Peter to his wife. "The difference of our ages is the square of our son Steve's age, and the difference of their squares is the cube of that."

What is Steve's age?

Solution.   Let Peter's age be P, let his wife's age be W, and let Steve's age be S. Then

$$P - W = S^2.$$

The trick with this problem is to discover the meaning of the word "that". The only interpretation that leads to a meaningful solution is "the square of Steve's age".
Therefore,

$$P^2 - W^2 = (S^2)^3 = S^6.$$

The left hand side of the above factors into $(P+W)(P-W)$, from which we can conclude that

$$P + W = S^4,$$

since

$$P - W = S^2.$$

These two results taken together imply that

$$P = \frac{1}{2}(S^4 + S^2) \quad \text{and} \quad W = \frac{1}{2}(S^4 - S^2).$$

Notice that P and W are increasing functions of S. One can verify that S cannot reasonably be larger than 3 or smaller than 3, so that $S = 3$ is the only reasonable solution to the problem.

(194) *Problem.* The natural number N, in decimal form, can be concatenated on the right of any natural number M, in decimal form, to produce a third natural number $M\widehat{\ }N$ in decimal form. For example, if 2 is concatenated on the right side of 35 the number 352 is produced. We call N a "magic number" if N is always a divisor of $M\widehat{\ }N$. Among the natural numbers less than 130, how many "magic numbers" are there, and what are they?

Solution. If N has k digits, then the definition of a magic number stated above means that N is magic if

$$N \text{ evenly divides } 10^k M + N, \text{ for all } M,$$
$$\text{or} \quad N \text{ evenly divides } 10^k M, \quad \text{ for all } M.$$

Since this condition must hold for all natural numbers M, we could clearly choose one which has no factors in common with N. This would imply that

$$N \text{ evenly divides } 10^k.$$

Since we are interested only in those numbers that are less than 130, we need only consider 1, 2 or 3 digit numbers as candidates. The 1-digit divisors of 10 are 1, 2, and 5; the 2-digit divisors of 100 are 10, 20, 25, and 50; the 3-digit divisors of 1000 which are less than 130 are 100 and 125.

Thus, there are 9 magic numbers less than 130, namely 1, 2, 5, 10, 20, 25, 50, 100, and 125.

(195) *Problem.* An eccentric artist says that the best canvasses have the same area as their perimeter. Let us not argue whether such sizes increase the onlooker's appreciation, but only try to find what sides (in integers only) a rectangle must have if its area and perimeter are equal.

Solution. Let a and b be the two dimensions and suppose that b is at least as large as a. Then

$$2a + 2b = ab,$$

from which we can compute that

$$b = \frac{2a}{a - 2}.$$

From this expression it is clear that $a > 2$ (otherwise b would be negative). By a similar argument we can conclude that $b > 2$. If we now use the assumption that $a < b$, we obtain the quadratic inequality

$$a^2 - 4a < 0,$$

from which we can conclude that $a < 4$. The only possibilities are $a = 3$ and $a = 4$, which yield the corresponding values $b = 6$ and $b = 4$. Thus, the solutions are 4×4 and 3×6.

(196) *Problem.*   In voting on a Canadian constitutional amendment, each province can either vote YES, vote NO or ABSTAIN.

In order for the amendment to pass, 7 or more of the 10 provinces must vote YES, and at least one of Ontario and Quebec must be among the YES votes.

How many different ways are there for the provinces to cast their votes so that the amendment is passed?

Solution.   The passing of the amendment requires that 7, 8, 9, or 10 provinces vote yes.

In the cases of nine or ten provinces voting yes, it is known that at least one of Ontario or Quebec voted yes.

Therefore, this can happen in $\binom{10}{10} + \binom{10}{9} = 1 + 10 = 11$ ways.

There are $\binom{10}{8} = 45$ ways that eight provinces could vote yes, though one of these would not result in the passing of the amendment as it would include Ontario and Quebec voting no. Hence, this gives an additional 44 possibilities.

Finally, there are $\binom{8}{6} \times \binom{2}{1} + \binom{8}{5} \times \binom{2}{2}$ ways of exactly seven provinces voting yes, including exactly one of Ontario and Quebec and six others, or both Ontario and Quebec, along with five others. This provides another $(28)(2) + (56)(1) = 112$ ways.

The number of possible ways that the vote can be cast to pass the amendment is then $1 + 10 + 44 + 112 = 167$.

(197) *Problem.*   A number consists of 22 digits and the last digit is 7. When the 7 is moved to the front of the number from the end, the number is increased to seven times its original value. Find the number.

Solution.   Let x be the number in question. Then $x = 10n + 7$, since x ends in a 7. Then $7x = 7 \cdot 10^{21} + n$. Solving this pair of equations simultaneously, we obtain

$$70x = x - 7 + 7 \cdot 10^{22}$$

from which we can obtain

$$69x = 7(10^{22} - 1).$$

Therefore,

$$x = \frac{7 \cdot 3{,}333{,}333{,}333{,}333{,}333{,}333{,}333}{23},$$

which yields $x = 1{,}014{,}492{,}753{,}623{,}188{,}405{,}797$.

(198) *Problem.*   Find all five-digit numbers which have their digits completely reversed by multiplying by 9.

Solution.   Let n be the number in question. Then

$$n = 10000a + 1000b + 100c + 10d + e,$$

$$\text{and} \quad 9n = 10000e + 1000d + 100c + 10b + a.$$

Summing, we obtain:

$$10n = 10000(a + e) + 1000(b + d) + 100(2c) + 10(b + d) + (a + e).$$

But,

$$10n = 100000a + 10000b + 1000c + 100d + 10e.$$

Clearly, by considering the units digit in the above we get $a + e = 10$, but since $9n$ is still a 5-digit number, we must have $a = 1$ and $e = 9$. Thus, by considering the tens digit of $10n$ above we see that $b + d + 1 = e = 9$ or $b + d + 1 = e + 10 = 19$, that is, $b + d = 8$ or $b + d = 18$. By considering the ten-thousands digit in $10n$, we conclude that $b = 0$ or $b = 1$, since $a + e = 10$. Clearly, the only possibility is that $b + d = 8$ and $b = 0$, which implies that $d = 8$. It is now an easy matter to show that $c = 9$.

Therefore, $n = 10989$ is the only 5-digit which simply reverses its digits on multiplication by 9.

(199) *Problem.*   Having been dragged into the crowded store by their wives to help choose Christmas cards, Les and Len were very glad to complete their purchases and escape to a nearby restaurant for refreshment. As they sat sipping their coffee, they started comparing their cards. Quite by chance it turned out that each couple had bought twenty; Les had bought three more than Ann, and Elsie had bought only four herself. It may seem quite irrelevant, but which of the two girls was married to Len?

Solution.   Since 20 is an even number, the number of cards purchased by each husband and wife are either both even or both odd. This means that the difference in the number of cards that they purchase must be an even number.

Since the difference between the number of cards purchased by Les and by Ann is 3, an odd number, we see that Les and Ann are not married to each other.

Thus, Ann is married to Len, and Elsie to Les. This solves the problem stated. It is now, however, a simple matter to work out individual purchases: Elsie 4, Les 16, Ann 13, and Len 7.

(200) *Problem.* Show that if p is any prime number different from 2 and 5 then there is a multiple of p which is written in base 10 as a string of ones. Can you generalize this?

Solution. Suppose that the statement in question is false. We must show that this supposition leads to a contradiction. Let p be a prime such that no multiple of it is composed entirely of ones when written in base 10 (the existence of p is guaranteed by our supposition).

Let n_k be a string of k ones and we will now start a long division of p into n_k for all values of k from 1 to p. At each step of the long division we will obtain a remainder r such that $0 < r < p$.

Since there are p ones (and thus p remainders) and at most $p - 2$ values for these remainders, we must eventually obtain two remainders having the same value, say $r_s = r_t$, where r_k is the remainder when n_k is divided by p.

Since one of s and t must be larger, let us assume that $s > t$. Let $n = n_s - n_t$. This is a number consisting of $s - t$ ones followed by t zeroes, which implies that $n = n_{s-t} \times 10^t$. However, n also is a multiple of p, since n_s and n_t leave the same remainder on division by p.

Since p is a prime different from 2 and 5, we may conclude that p divides evenly into n_{s-t}, which contradicts our original supposition. Therefore, the statement in the problem is true.

(201) *Problem.* All the 2-digit numbers from 19 to 80 are written in a row. The result is read as a single integer $19202122\ldots787980$. Is this integer divisible by 1980?

Solution. Let n be the number defined in the problem. Since $1980 = 20 \cdot 9 \cdot 11$, and since all three factors are relatively prime (that is, have no factors in common, except 1), it is sufficient to determine if n is divisible by each of 20, 9, and 11. To determine if n is divisible by 20 we need only examine its last two digits, namely 80. Since 80 is divisible by 20, so is n.

To determine if n is divisible by 9, it is sufficient to determine if the sum of its digits is divisible by 9. The sum of its digits is

$$1 + 9 + 8 + 0 + 10(2 + 3 + 4 + 5 + 6 + 7)$$
$$+ 6(0 + 1 + 2 + 3 + 4 + 5 + 6 + 7 + 8 + 9) = 558$$
$$= 9(62),$$

where the two sums with parentheses correspond to the tens and units digits of the integers from 20 to 79 inclusive. As you can see, this sum is divisible by 9, whence so is n. To determine if n is divisible by 11, it is sufficient to determine if the sum of its digits with alternating signs is divisible by 11. This sum is

$$1 - 9 + 8 - 0 + 10(2 + 3 + 4 + 5 + 6 + 7)$$
$$- 6(0 + 1 + 2 + 3 + 4 + 5 + 6 + 7 + 8 + 9) = 0,$$

which is clearly divisible by 11. Thus, n is also divisible by 11. Therefore, n is divisible by 1980.

(202) *Problem.* Find the largest even natural number which cannot be expressed as the sum of two composite odd natural numbers.

Solution. Let n be an even natural number. Then each of the following expresses n as the sum of two odd numbers:

$$n = (n - 15) + 15,\ n = (n - 25) + 25,\ n = (n - 35) + 35.$$

Because 15, 25 and 35 all leave different remainders on division by 3, so also do the numbers $n - 15$, $n - 25$ and $n - 35$. Thus, one of them is evenly divisible by 3. Now, for $n > 38$, all three of these values exceed 3 and thus, at least one of them is composite (since one of them is divisible by 3). Since 15, 25 and 35 are all composite, we have expressed n as the sum of two odd composite numbers. It can be verified that 38 cannot be so expressed. Therefore, 38 is the largest even natural number which cannot be expressed as the sum two odd composite numbers.

(203) *Problem.* All students at Adams High School and at Baker High School take a certain exam. The average scores for boys is 71 and 81 at Adams and Baker, respectively; for girls it is 76 and 90, respectively; and for boys and girls combined it is 74 and 84, respectively. If the average for boys at the two schools combined is 79, what is the average score for girls at the two schools combined?

Solution. Let b_1 and b_2 be the number of boys at Adams and Baker, respectively, and let g_1 and g_2 be the number of girls at

Adams and Baker, respectively. Similarly let M_1, M_2, F_1, and F_2 be the total scores for the boys and girls at Adams and Baker, respectively. Then

$$M_1 = 71b_1, \tag{1}$$
$$M_2 = 81b_2, \tag{2}$$
$$F_1 = 76g_1, \tag{3}$$
$$F_2 = 90g_2, \tag{4}$$
$$M_1 + F_1 = 74(b_1 + g_1), \tag{5}$$
$$M_2 + F_2 = 84(b_2 + g_2), \tag{6}$$
$$M_1 + M_2 = 79(b_1 + b_2). \tag{7}$$

Combining equations (1), (2) and (7) we get $b_2 = 4b_1$. Combining equations (1), (3) and (5) we have $3b_1 = 2g_1$, which implies that b_1 is a multiple of 2. Further combining equations (2), (4) and (6) we obtain $b_2 = 2g_2$. Now if we let $b_1 = 2k$ where k is some natural number, we must have $b_2 = 8k$, $g_1 = 3k$ and $g_2 = 4k$. To compute the average score among the girls at the combined schools, we evaluate

$$\frac{F_1 + F_2}{g_1 + g_2} = \frac{76g_1 + 90g_2}{g_1 + g_2} = \frac{76(3k) + 90(4k)}{3k + 4k} = 84.$$

Thus, the average we seek is 84.

(205) *Problem.*   A person cashes a cheque at the bank. By mistake, the teller pays the number of cents as dollars and the number of dollars as cents. The person spends \$3.50 before noticing the mistake, then on counting the money finds that there is exactly double the amount of the cheque. For what amount was the cheque made out? *Solution.*   Let a be the number of dollars in the cheque and b be the number of cents. Then the total amount of the cheque in cents was $100a + b$. The total amount received (in cents) was $100b + a$. Notice that both a and b are positive integers and must not exceed 99. After counting, it is discovered that

$$100b + a - 350 = 2(100a + b).$$

This equation simplifies to $199a = 98b - 350 = 14(7b - 25)$. Since the right hand side is divisible by 14, so is the left hand side, whence a is divisible by 14. Therefore, let $a = 14c$, where c is also a positive integer. Now the equation can be rewritten as

$$7b - 199c = 25.$$

There are an infinite number of solutions to this equation in two variables. One such solution is $b = 32$ and $c = 1$. For integer solutions b and c, the general solution is

$$b = 32 + 199k \qquad \text{and} \qquad c = 1 + 7k$$

for all integers k. Since b is a positive integer not exceeding 99, we must conclude that $k = 0$. Therefore, $b = 32$, $c = 1$ and $a = 14$. Thus, the amount of the original cheque was \$14.32.

(205) *Problem.* Suppose that m and n are positive integers and that the decimal expansion of the rational number m/n has a repetend of 4356. Prove that n is evenly divisible by 101.

Solution. Let a be the rational number m/n. Suppose that there are k digits following the decimal place before the expansion repeats. Then note that both $10^k a$ and $10^{k+4} a$ have the same sequence of digits following the decimal place, namely $43564356\ldots$. Now let $b = 10^{k+4}a - 10^k a$. Clearly, b is an integer. If we now let c be the integer part of $10^k a$, then the integer part of $10^{k+4}a$ is $10^4 c + 4356$. Thus $b = 10^4 c + 4356 - c = 9999c + 4356$. From the definition of b above we also have $10^k(9999)a = b$. Putting it all together, we get

$$\frac{m}{n} = a = \frac{b}{10^k(9999)} = \frac{9999c + 4356}{10^k \cdot 9 \cdot 11 \cdot 101},$$

which simplifies to

$$m \cdot 10^k \cdot 9 \cdot 11 \cdot 101 = n(9999c + 4356).$$

Clearly 101 evenly divides the right hand side of the above equation. Now for any value of c, 101 cannot evenly divide $9999c + 4356$, since it evenly divides 9999 and does not evenly divide 4356. Since 101 is a prime number, we may conclude that it evenly divides n.

(206) *Problem.* How many primes among the positive integers, written as usual in base 10, are such that their digits are alternating 1's and 0's, beginning and ending with 1?

(A prime number p is a positive integer other than 1 which has exactly two divisors, namely p and 1.)

Solution. Let N_k be the number $101010\ldots101$ with exactly k digits equal to 0. If $k = 1$, we get $N_k = 101$, which is prime. Now let us consider $k \geq 2$. First let k be odd. Then N_k can be seen as successive copies of 101 separated by 0's, which clearly implies that 101 divides evenly into N_k, whence N_k is not prime. for example, $k = 3$

$$10101010101 = 101\,0\,101\,0\,101 = 101 \times 100010001.$$

Let us suppose instead that k is even. Observe that $11N_k$ is a string of $2k + 2$ ones. This can always be factored into a product of two numbers r and s, where r is a string of $k + 1$ ones and s is a number having k zeroes with a single one added to both ends.

This implies that $11N_k = rs$. Since 11 is a prime number, it must divide evenly into either r or s. In either case, the other of the two numbers r and s must evenly divide into N_k.

Since neither r nor s can equal N_k or 1, we see that again N_k cannot be prime. Indeed we can conclude that N_k is prime only if $k = 1$, which means that there is only number of the required form which is prime.

(207) *Problem.* In the sum below the powers of 3 have been written under each other, but with the final digit of each number displaced (or shifted) one place to the right compared to the preceding number. Assuming that the powers of 3 are added up forever in this manner, what will be the resulting string of digits.

1

3

9

27

81

243

729

$\ddots$

Solution. One of the easiest ways to solve this problem is to pretend there is a decimal point immediately in front of the number 1 in the first row. In this case the nth row corresponds to the value $\frac{3^{n-1}}{10^n}$. Thus, we would be looking at summing up the geometric series:

$$S = \frac{1}{10} + \frac{3}{100} + \frac{9}{1000} + \frac{81}{10000} + \cdots,$$

which has initial term $\frac{1}{10}$ and common ratio $\frac{3}{10}$. The sum for such a geometric series is

$$S = \frac{\dfrac{1}{10}}{1 - \dfrac{3}{10}} = \frac{1}{7}.$$

Thus, the digits of the sum are simply the repeating sequence $142857142857\ldots$ obtained from the decimal expansion of $\frac{1}{7}$.

(208) *Problem.* Let n be an even number which is divisible by a prime bigger than $\sqrt{n}$. Show that n and n^3 cannot be expressed in the form $1 + (2k+1)(2k+3)$; that is, as one more than the product of two consecutive odd numbers, but that n^2 and n^4 can be so expressed.

Solution. Call a natural number "expressible" if it is of the form $1 + (2k+1)(2k+3)$ for some k. If n is even, then

$$n^2 = 1 + (n-1)(n+1) = 1 + (2k+1)(2k+3) \qquad \text{where } k = \frac{n-2}{2},$$

showing that n^2 is expressible. Since n^2 is even when n is even, n^4 is also expressible. On the other hand suppose that n is expressible. Then $n = 4k^2 + 8k + 4 = 4(k+1)^2$ showing that the largest prime divisor of n cannot exceed either 2 or $k+1$, which is a contradiction since there is a prime divisor which exceeds $\sqrt{n} = 2(k+1)$. Now suppose that n^3 is expressible. Then $n^3 = (2(k+1))^2$, implying that n is a perfect 6th power. Hence, the largest prime divisor of n^3 (and thus, the largest prime divisor of n) cannot exceed $(n^3)^{1/6}$ or $\sqrt{n}$, which is again a contradiction.

(209) *Problem.* Solve the following cryptarithm:

$$\frac{\text{EVE}}{\text{DID}} = .\text{TALKTALKTALKTALK}\ldots$$

where each letter represents a different decimal digit, and the expression on the right is a repeating decimal fraction with period 4. The fraction on the left hand side has been reduced to its lowest terms. (This is one of the oldest such cryptarithms and one of the best, and is not intended to be sexist!)

Solution. First of all let the fraction be r. Then note that $10000r = \text{TALK} + r$, which means that

$$r = \frac{\text{EVE}}{\text{DID}} = \frac{\text{TALK}}{9999}$$

or $9999(\text{EVE}) = (\text{TALK})(\text{DID}).$

Thus, the number DID must be a divisor of $9999 = 3^2 \cdot 11 \cdot 101$. The only such divisors are 101, 303, and 909.

Suppose first that DID = 101. Then TALK = 99(EVE). EVE cannot be 101 since that is taken, and any higher value gives a 5-digit product. Thus DID cannot be 101. Next suppose that DID = 909. Then TALK = 11(EVE). In this case the last digit of 11(EVE) would have to be E, but it is K. Therefore DID cannot be 909 either. So if there is any solution we must have DID = 303. Then TALK = 33(EVE), which means that EVE is smaller than 303. There are 14 possibilities for EVE since the digits 0 and 3 must be avoided, namely, 121, 141, 151, 161, 171, 181, 191, 212, 242, 252, 262, 272, 282, and 292. Since EVE and DID have no common factors, we can eliminate all multiples of 3 from this list, which gives us 10 possibilities: 121, 151, 161, 181, 191, 212, 242, 262, 272, and 292. At this point we need to explore the decimal expansions of each number in this list divided by DID = 303. Only EVE = 242 has a decimal expansion where each of the four repeated digits is different from the digits of both DID and EVE.

Thus, our solution is

$$\frac{242}{303} = .798679867986\ldots$$

(210) *Problem.* Consider the multiplication $d \times dd \times ddd$, where $d < b-1$ is a non-zero digit in base b and the product (base b) has 6 digits, all less than $b - 1$ as well. Suppose that, when d and the digits of the product are all increased by 1, the multiplication is still true. Find the smallest base b in which this can happen.

Solution. We have (in base b):

$$d^3(1)(11)(111) + 111111 = (d+1)^3(1)(11)(111)$$

or

$$111111 = (11)(111)(3d^2 + 3d + 1),$$

which simplifies to

$$1001 = 11(3d^2 + 3d + 1),$$

since $111111 = (1001)(111)$ in any base. If we now evaluate this last equation we get

$$b^3 + 1 = (b+1)(3d^2 + 3d + 1)$$
$$b^2 - b + 1 = 3d^2 + 3d + 1,$$

or

$$b(b - 1) = 3d(d + 1).$$

This equation is clearly satisfied by $d = 1$, $b = 3$ which we reject since the product$[1 \cdot 11 \cdot 111 = 1221_3]$ has less than 6 digits. The lowest base which satisfies the problem is $b = 10$, and then $d = 5$, where

$$5(55)(555) = 152625, \qquad 6(66)(666) = 263736.$$

(211) *Problem.* An n-digit integer is *cute* if its n digits are an arrangement of the set $\{1, 2, \ldots, n\}$ and its first k digits form an integer that is divisible by k, for $k = 1, 2, \ldots, n$. For example, 321 is a cute 3-digit integer because 1 divides 3, 2 divides 32, and 3 divides 321. Determine all cute 6-digit positive integers.

Note: we are always dealing with base 10 notation here.

Solution. Since the first 5 digits of a cute number must form an integer divisible by 5, the digit 5 must occupy the fifth position. Since a number must be even if it is divisible by an even digit, there must be even digits in the second, fourth, and sixth positions.

Thus, 1 and 3 must occupy the first and third positions (not necessarily in that order). Divisibility by 6 is ensured since the sixth digit is even and the sum of the digits id 21 (a mulitple of 3).

When a number is divisible by 3, the sum of its digits is also divisible by 3.

Therefore, the second digit must be 2, since $1 + 2 + 3 = 6$ is divisible by 3, but $1 + 4 + 3 = 8$ and $1 + 6 + 3 = 10$ are not. Since the first four digits must form an integer that is divisible by 4, we see that digits three and four of our cute number must form a 2-digit number which is divisible by 4, which implies that the fourth digit is 6 (since both 16 and 36 are divisible by 4, but neither 14 nor 34 are). This leaves us with two possible cute six-digit numbers: 123654 and 321654.

(212) *Problem.* What is the smallest positive integer that can be expressed as the sum of nine consecutive positive integers, the sum of ten consecutive positive integers, and the sum of eleven consecutive positive integers?

Solution. Let n be the number we seek. Any set of 9 consecutive integers will sum to 9 times the middle number, and hence, n must be divisible by 9. A similar argument holds for 11, thus, making n divisible by 11. In the case of 10 consecutive integers, the sum equals 5 times the sum of the middle pair of numbers (as there are

really 5 such pairings of smallest with largest, second smallest with
second largest, and so on).

Therefore, n must be divisible by 5. Since 5, 9, and 11 have no
common factors, we require that n be a multiple of their product
$5 \times 9 \times 11 = 495$. We can check that 495 works by taking $495/9 = 55$
and $495/11 = 45$ as the middle numbers in their respective sums.
Further, $495/5 = 99 = 49 + 50$, thus, making them the middle pair
of numbers in the other sum. Thus, we have

$$
\begin{aligned}
495 &= 51 + 52 + 53 + 54 + 55 \\
&\quad + 56 + 57 + 58 + 59 \\
&= 45 + 46 + 47 + 48 + 49 \\
&\quad + 50 + 51 + 52 + 53 + 54 \\
&= 40 + 41 + 42 + 43 + 44 \\
&\quad + 45 + 46 + 47 + 48 + 49 + 50.
\end{aligned}
$$

(213) *Problem.*   What fraction has the smallest denominator in the
open interval $\left(\frac{19}{94}, \frac{17}{76}\right)$?

Solution.   Let a/b be the fraction in the given interval with the
smallest denominator. Then we clearly have

$$
\frac{19}{94} < \frac{a}{b} < \frac{17}{76}
$$

$$
\text{or} \qquad 4 + \frac{18}{19} = \frac{94}{19} > \frac{b}{a} > \frac{76}{17} = 4 + \frac{8}{17}.
$$

If we set $b = 4a + c$, then $0 < c < a$ and we have

$$
\frac{18}{19} > \frac{c}{a} > \frac{8}{17}
$$

$$
\text{or} \qquad 2 + \frac{1}{8} = \frac{17}{8} > \frac{a}{c} > \frac{19}{18} = 1 + \frac{1}{18}.
$$

In order to find the smallest value of $b = 4a + c$ which will work, we
must find the smallest value of a and c. From the last inequality
above we see that the smallest values of a and c with $0 < c < a$
which work is $a = 2$ and $c = 1$. This implies that $b = 9$ and the
fraction we seek is $\frac{2}{9}$, which can be seen to fall in the interval.

(214) *Problem.* A man specified in his will that his horses should be divided so that his eldest child would get $\frac{1}{2}$ of his horses, his middle child would get $\frac{1}{4}$, and his youngest child would get $\frac{1}{6}$. When he died he owned 11 horses. His lawyers were in a quandary as to how to comply with the his wish since a horse divided is not much of a horse at all! A relative of the heirs solved the problem by loaning them one of his own horses, so that there were 12 in total. The eldest took her 6, the middle child her 3, and the youngest his 2. The one horse remaining was returned to the relative.

How many sets of numbers are there for the number n of horses and the three fractions (reciprocals of distinct integers a, b, and c) such that by borrowing a horse, the heirs could comply with the will, each receive a whole number of horses, and return the borrowed horse?

Solution. We are looking for the number of solutions in integers to the equation:

$$\frac{n}{n+1} = \frac{1}{a} + \frac{1}{b} + \frac{1}{c}$$

where $1 < a < b < c$. After borrowing the horse, there are $n+1$ horses and each heir must receive a whole number of horses. Thus, $n+1$ must be a multiple of the least common multiple of a, b, and c. On the other hand, the left hand side of the above equation is a reduced fraction and the right hand side could be expressed as a fraction with the least common multiple as the denominator, which implies that $n+1$ divides evenly into the least common multiple of a, b, and c. Therefore, we must have $n+1$ equal to the least common multiple of a, b, and c.

If $a \geq 3$, then the least common multiple of a, b, and c is at least 12. This implies that $\frac{n}{n+1} \geq \frac{11}{12}$. But then,

$$\frac{1}{a} + \frac{1}{b} + \frac{1}{c} \leq \frac{1}{3} + \frac{1}{4} + \frac{1}{5} = \frac{47}{60} < \frac{11}{12} \leq \frac{n}{n+1},$$

which is impossible. Thus, $a = 2$.

If $b \geq 5$, then the least common multiple of a, b, and c is at least 10. This implies that $\frac{n}{n+1} \geq \frac{9}{10}$. But then,

$$\frac{1}{a} + \frac{1}{b} + \frac{1}{c} \leq \frac{1}{2} + \frac{1}{5} + \frac{1}{6} = \frac{52}{60} < \frac{9}{10} \leq \frac{n}{n+1},$$

which is impossible. Thus, $b = 3$ or 4.

If we consider the case $a = 2$ and $b = 4$, we see that the equation above can be rearranged to yield:

$$(c - 4)(n - 3) = 16,$$

with $c > 4$. We now simply let $c - 4$ run through the factors of 16 giving us values of (c, n) equal to $(5, 19)$, $(6, 11)$, $(8, 7)$, $(12, 5)$, and $(20, 4)$. In the last two the least common multiple is not equal to $n + 1$ (since $n + 1$ is not even a multiple of c).

If we consider the case $a = 2$ and $b = 3$, we see that the equation can be rearranged to yield:

$$(c - 6)(n - 5) = 36,$$

with $c > 3$ and $c \neq 6$. It is quickly seen that $c = 4$ and $c = 5$ yield negative values for n and thus, are impossible. Therefore we have $c > 6$. We now simply let $c - 6$ run through the factors of 36 giving us values of (c, n) equal to $(7, 41)$, $(8, 23)$, $(9, 17)$, $(10, 14)$, $(12, 11)$, $(15, 9)$, $(18, 8)$, $(24, 7)$, and $(42, 6)$. Of these, only $(7, 41)$, $(8, 23)$, $(9, 17)$, and $(12, 11)$ are possible since as above $n + 1$ must be a multiple of c.

Thus there are 7 possible values of c and n:

a	b	c	n
2	3	7	41
2	3	8	23
2	3	9	17
2	3	12	11
2	4	5	19
2	4	6	11
2	4	8	7

(215) *Problem.*   I was at a restaurant the other day. The bill came, and I wanted to give the waiter a whole number of dollars, with the difference between what I give him and the bill being the tip. I always like to tip between 10 and 15 percent of the bill. But if I gave him a certain number of dollars, the tip would have been **less** than 10% of the bill, and if instead I gave him one dollar more, the tip would have been **more** than 15% of the bill. What was the largest possible amount of the bill?

Solution. Let n be the amount of the bill in cents. Let m be the whole number of dollars which would have included too small a tip. Then $m + 1$ dollars would have included too large a tip. Thus, in cents we have

$$100m < 1.1n, \tag{1}$$

$$100(m + 1) > 1.15n. \tag{2}$$

These two inequalities immediately imply that $0.05n < 100$ or $n < 2000$. By combining this with (1) we see that $m \le 21$. Together with (2), this further implies that $n \le 1913$. This value of n actually satisfies both inequalities and is thus, the largest such value. Therefore, the largest amount of the bill is \$19.13.

(216) *Problem.* A rectangle is drawn on graph paper as shown below and the border cells are shaded. In this case there are more border cells than interior cells. Is it possible to draw a rectangle of proportions such that the border (one cell wide) contains the same number of cells as the interior? If so, how many such rectangles are there and what are the possible dimensions?

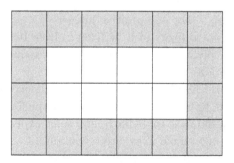

Solution. Let x and y be the number of cells in the two sides of the rectangle. We may assume that $x \ge y$. The total number of cells in the rectangle is xy.

We want the number of cells in the border to be $\frac{1}{2}xy$. But we have exactly $2x + 2y - 4$ border cells. Setting these two values equal to each other and rearranging we get

$$(x - 4)(y - 4) = 8.$$

Thus, $x - 4$ and $y - 4$ must both be positive integer divisors of 8. Since $x \ge y$ this leaves us with the two solutions (x, y) equal to $(12, 5)$ and $(8, 6)$.

(217) *Problem.* Four numbers are given. Three of the numbers are chosen, their average is taken, and the fourth number is added to this average. If all possible ways of doing this give 17, 21, 23, and 29, what were the original numbers?

Solution. Let a, b, c, and d be the four numbers in question. Then we have:

$$\frac{a+b+c}{3} + d = 17$$

$$\text{or} \quad a + b + c + 3d = 51.$$

A similar calculation will also yield:

$$a + b + 3c + d = 63$$

$$a + 3b + c + d = 69$$

$$3a + b + c + d = 87.$$

Adding all four equations gives us

$$6a + 6b + 6c + 6d = 270,$$

from which we have $a + b + c + d = 45$. We can now combine this with each of the above four equations to obtain the values $a = 21$, $b = 12$, $c = 9$, and $d = 3$.

(218) *Problem.* For any positive integer n, let $P(n)$ be the product of the non-zero digits of n, written in base 10. Call n "prodigitious" if $P(n)$ divides evenly into n. Find a sequence of 13 consecutive prodigitious positive integers.

Solution. We may not allow any digit except 0 or 1 to occur in the base 10 representation of n before the final digit (otherwise this digit would have to divide n, and some experimentation shows that this is impossible if we are to obtain 13 consecutive prodigitious positive integers). Since an integer ending in 14 in base 10 is never evenly divisible by 4, we must be looking at the last two digits in the range from 00 to 13. Since a number is evenly divisible by 3 if and only if the sum of its digits is divisible by 3, we see that numbers ending in 03 and 13 cannot both be present in the string of 13 consecutive prodigitious integers. Thus, they must end in 00 through to 12, but cannot be this set of integers since they must all be <u>positive</u>. Thus we must put a string of 1s and/or 0s in front of these final 2 digits. Since this string must be divisible by 9 its digits must sum to a multiple of 9, and since it must also be divisible by 7, a multiple of 1001 will also work. One candidate for such a string is 18 ones in a row. This would yield the solution:

111111111111111111100
111111111111111111101
111111111111111111102
111111111111111111103
111111111111111111104
111111111111111111105
111111111111111111106
111111111111111111107
111111111111111111108
111111111111111111109
111111111111111111110
111111111111111111111
111111111111111111112

and the reader can verify that all are prodigitious.

(219) *Problem.* Walter shook his head. "I closed down last week, and I'm broke," he said. "The last day we took in precisely $5.33, and that was the end."

"I'm sorry," Andy was shocked. "You started well, with good sales the first day and people standing in line."

"Maybe too well. Every day after that we took in less," Walter told him. "It was uncanny, like a curse. The second day we took in just one cent more than two-thirds of the first day's take, the third day two cents more than two-thirds of the second day, the fourth day three cents more than two-thirds of the third day, and so on. Each day another penny more than two-thirds of the previous day."

He had a real problem, and so will you if you want to solve this! How much had Walter taken in the first day, and how many days was the business open?

Solution. Take the first day as day number 0 and let its sales be x cents. Then on successive days sales in cents were:

$$0: \qquad x$$

$$1: \qquad \tfrac{2}{3}x + 1$$

$$2: \qquad \tfrac{2}{3}\left(\tfrac{2}{3}x + 1\right) + 2 = \left(\tfrac{2}{3}\right)^2 x + \tfrac{2}{3} + 2$$

$$3: \qquad \tfrac{2}{3}\left(\left(\tfrac{2}{3}\right)^2 x + \tfrac{2}{3} + 2\right) + 3 = \left(\tfrac{2}{3}\right)^3 x + \left(\tfrac{2}{3}\right)^2 + 2\left(\tfrac{2}{3}\right) + 3$$

$$4: \tfrac{2}{3}\left(\left(\tfrac{2}{3}\right)^3 x + \left(\tfrac{2}{3}\right)^2 + 2\left(\tfrac{2}{3}\right) + 3\right) + 4 = \left(\tfrac{2}{3}\right)^4 x + \left(\tfrac{2}{3}\right)^3 + 2\left(\tfrac{2}{3}\right)^2$$
$$+ 3\left(\tfrac{2}{3}\right) + 4$$

$$\vdots$$

By mathematical induction, sales for day number n can be arranged as

$$\left(\tfrac{2}{3}\right)^n x + \sum_{i=1}^{n} i \left(\tfrac{2}{3}\right)^{n-i} = \left(\tfrac{2}{3}\right)^n x + \sum_{i=1}^{n}\sum_{k=1}^{i} \left(\tfrac{2}{3}\right)^{k-1}$$

$$= \left(\tfrac{2}{3}\right)^n x + \sum_{i=1}^{n} \frac{1 - \left(\tfrac{2}{3}\right)^i}{1 - \tfrac{2}{3}}$$

$$= \left(\tfrac{2}{3}\right)^n x + 3\sum_{i=1}^{n}\left(1 - \left(\tfrac{2}{3}\right)^i\right)$$

$$= \left(\tfrac{2}{3}\right)^n x + 3\sum_{i=1}^{n} 1 - 3\sum_{i=1}^{n} \left(\tfrac{2}{3}\right)^i$$

$$= \left(\tfrac{2}{3}\right)^n x + 3n - 3\left(\frac{1 - \left(\tfrac{2}{3}\right)^{n+1}}{1 - \tfrac{2}{3}}\right)$$

$$= \left(\tfrac{2}{3}\right)^n x + 3n - 9 + 9\left(\tfrac{2}{3}\right)^{n+1}$$

$$= \left(\tfrac{2}{3}\right)^n (x + 6) + 3n - 9$$

$$= \frac{2^n(x+6)}{3^n} + 3n - 9.$$

If n was the last day of business then this expression must equal 533 cents. Thus $2^n(x+6)/3^n + 3n = 539$, which implies that $x+6 = 3^n k$ for some integer k, further implying that $2^n k = 539 - 3n$. From the wording of the problem we have $n > 3$. Also, since $2^{10} = 1024 > 539$, we have $n < 10$. And $539 - 3n$ must be even which means that n is odd. Thus, $n = 5$, 7, or 9. Checking these values we only get an integer value for k when $n = 9$. Therefore, $n = 9$, $k = 1$ and $x = 19677$. The first day's sales were \$196.77, and they were open for business for 10 working days.

(220) *Problem.*   For a certain positive composite integer x, when the fraction $(60 - x)/120$ is reduced to lowest terms, the sum of the numerator and denominator exceeds 120. Determine x. (An integer is called *composite* if it is evenly divisible by an integer which is different from itself and different from 1.)

Solution.   Let $(60 - x)/120 = a/b$ in lowest terms.

If $x \geq 60$, then $60 - x \leq 0$. Then $a + b \leq 180 - x = 120 + (60 - x) \leq 120$. Hence, $x \leq 60$.

Now, if x is a multiple of 2, then we have $x = 2y$ for some integer y, so that $(60 - x)/120 = (30 - y)/60$, which implies that $a + b \leq$

$90 - y < 120$. Thus, x is not a multiple of 2.

If x is a multiple of 3, then $x = 3y$ for some integer y, so that we have $(60 - x)/120 = (20 - y)/40$, which implies that $a + b \leq 60 - y < 120$. Thus, x is not a multiple of 3.

Finally, if x is a multiple of 5, then $x = 5y$ for some integer y, so that $(60 - x)/120 = (12 - y)/24$, which implies that $a + b \leq 36 - y < 120$. Thus, x is not a multiple of 5.

We have shown that the composite integer x is at most 60 and is not a multiple of 2, 3, or 5. The only such number x is 49. We check $x = 49$: $(60 - x)/120 = 11/120$, which is in lowest terms, and $11 + 120 > 120$. Therefore, the answer is $x = 49$.

(221) *Problem.* Let n be a natural number. A cube of side n can be split into 2000 cubes. The sides of these cubes are also natural numbers. Determine the minimum possible value of n.

Solution. Since $12^3 = 1728 < 2000$, we see that $n \geq 13$. To show that $n = 13$ is the answer, we need only express 13^3 as the sum of 2000 cubes of positive integers. One such representation is

$$2197 = 13^3 = 1(5^3) + 2(3^3) + 3(2^3) + 1994(1^3),$$

since $2000 = 1 + 2 + 3 + 1994$. Let us see how to find all the possible decompositions of 13^3 into 2000 cubes of positive integers.

If we divide the cube of side 13 into cubes of side 1 we have 2197 individual cubes, which is 197 more than we want. We thus need to glue some of these small cubes into larger ones, thereby reducing the total number of cubes. Now replacing 8 cubes of side 1 with 1 cube of side 2 reduces the total by 7; replacing 27 cubes of side 1 with 1 cube of side 3 reduces the total by 26; replacing 64 cubes of side 1 with 1 cube of side 4 reduces the total by 63; and replacing 125 cubes of side 1 with 1 cube of side 5 reduces the total by 124. We do not need to consider any larger cubes since $6^3 - 1 = 216 - 1 = 215$ is larger than the total reduction needed (that is, larger than 197). It remains for us to find an integer solution to the equation:

$$7x + 26y + 63z + 124w = 197.$$

Since 7 evenly divides 63, we may initially ignore the term $63z$, and only consider it later for solutions with $x \geq 9$. Thus, we will consider the equation:

$$7x + 26y + 124w = 197.$$

Clearly $w \leq 1$. Let us first consider $w = 0$. Then $7x + 26y = 197$. Since the integer part of $197/26$ is 7, we need only examine the 8 possible values of y from 0 to 7.

We get only the solution $(x, y) = (17, 3)$, which in light of our earlier comment can be converted into the solutions $(x, y, z, w) = (17, 3, 0, 0)$ and $(8, 3, 1, 0)$. It is easy to verify that cubes of the appropriate sizes can all be located within a cube of side 13, which shows that all are genuine solutions.

Now consider $w = 1$. We are left with the equation $7x + 26y = 73$, for which the only solution is $(x, y) = (3, 2)$, which yields the solution $(x, y, z, w) = (3, 2, 0, 1)$ (which was the original one mentioned at the top), which is realizable since the smaller cubes can indeed be fitted inside a cube of side 13.

(222) *Problem.* If a, b, and c are positive integers such that $a \mid b^c$, show that $a \mid b^a$. (Note – when m and n are integers, we write $m \mid n$ if there is an integer k such that $n = mk$, in which case we say "m divides n".)

Solution. The proof is by contradiction. Suppose that the claim is false. Then there exists a prime p and a positive integer m such that p^m divides a, but not b^a. Since $p \mid a$, $a \mid b^c$ and p is a prime, we have $p \mid b$. Thus $p^a \mid b^a$, so that $a < m$. Then

$$2^a \leq p^a < p^m \leq a,$$

a contradiction. Thus, the claim must be true.

(223) *Problem.* Determine all primes p for which the system

$$p + 1 = 2x^2$$
$$p^2 + 1 = 2y^2$$

has a solution in integers x and y.

Solution. Clearly p is an odd prime. We will assume that both x and y are positive.

By subtracting the two given equations we have $p(p - 1) = 2(y - x)(y + x)$.

Since $p \neq 2$, we have either $p = y - x$ or $p = y + x$. In the former case, we would also have $p - 1 = 2(y + x) > y - x = p$, which is impossible.

Thus, $p = y + x$ and $p - 1 = 2(y - x)$. Eliminating p yields $y = 3x - 1$, which implies $p = x + y = 4x - 1$. Since we also have $p = 2x^2 - 1$, we get

$$2x^2 - 4x = 0$$

$$\text{or} \quad x(x - 2) = 0.$$

Since $x = 0$ is impossible, we have $x = 2$, whence $p = 7$ is the only solution.

(224) *Problem.* Find all integers n such that $n^2 - 11n + 63$ is a perfect square.

Solution. Let $n^2 - 11n + 63 = k^2$, where k is a non-negative integer. Then

$$4n^2 - 44n + 252 = 4k^2 \quad \Longrightarrow \quad (2n - 11)^2 + 131 = (2k)^2$$
$$\Longrightarrow \quad (2k)^2 - (2n - 11)^2 = 131$$
$$\Longrightarrow \quad (2k + 2n - 11)(2k - 2n + 11) = 131.$$

Notice that 131 is a prime. Since $2k+2n-11$ and $2k-2n+11$ are two integers whose sum, $4k$, is non-negative and whose product, 131, is positive, both of them are positive, implying that one of them is 1 and the other is 131. If $2k + 2n - 11 = 131$, then $2k - 2n + 11 = 1$, then subtracting yields $4n - 22 = 130$ or $n = 38$. On the other hand, if $2k + 2n - 11 = 1$ and $2k - 2n + 11 = 131$, then subtracting yields $4n - 22 = -130$ or $n = -27$. Hence, the only values yielding a perfect square for $n^2 - 11n + 63$ are $n = 38$ and -27.

(225) *Problem.* Solve, in integers, the equation

$$(x^2 - y^2)^2 = 1 + 16y.$$

Solution. First notice that if (x, y) is a solution, then so also is $(-x, y)$. Thus, we may assume that $x \geq 0$ for the purpose of solving and simply extend the solution set to negative values of x as appropriate.

The left hand side of the given equation is positive, which means that $y \geq 0$. In fact, $y = 0$ yields $x^4 = 1$, which means that in this case, we have only the solutions $(\pm 1, 0)$. Thus, from this point on, we will assume that $y > 0$. Also, we note that $x = 0$ implies that $y^4 = 16y + 1$, which has only one solution for $y > 0$: it is a value strictly between 2 and 3.

Thus, $x = 0$ produces no solutions. Therefore, from this point on, we will only consider $x > 0$ for the purpose of finding the solutions. The given equation can be factored to yield:

$$(x - y)^2(x + y)^2 = 1 + 16y.$$

Clearly, $x \neq y$. Then, $|x - y| \geq 1$. Suppose that $|x - y| \geq 2$. Then we have

$$1 + 16y = (x - y)^2(x + y)^2 \geq 4(x + y)^2$$
$$\geq 4(1 + y)^2 = 4y^2 + 8y + 4$$
$$> 4y^2 + 8y + 1$$
$$\text{or} \quad 16y > 4y^2 + 8y$$
$$2 > y.$$

Thus, $y = 1$, but in this case $16y + 1 = 17$, which is not a perfect square, and hence, not a solution. Therefore, we must have $|x-y| = 1$. That is, $x = y \pm 1$, which means that $x + y = 2y \pm 1$. Then

$$1 + 16y = (x - y)^2(x + y)^2 = (x + y)^2$$
$$= (2y \pm 1)^2 = 4y^2 \pm 4y + 1$$
$$\text{or} \quad 16y = 4y^2 \pm 4y$$
$$4 = y \pm 1,$$

which means that $y = 4 \mp 1$, that is, $y = 3$ (which corresponds to $x = y+1 = 4$) or $y = 5$ (which corresponds to $x = y-1 = 4$). Thus, our solutions for x, y positive are $(4, 3)$ and $(4, 5)$. In summary, the full set of integer solutions are $(x, y) = (\pm 1, 0)$, $(\pm 4, 3)$, and $(\pm 4, 5)$.

(226) *Problem.*   Find all values of a such that $x^3 - 6x^2 + 11x + a - 6 = 0$ has exactly three integer solutions.

Solution.   There are many different approaches available to solve this problem. We will only present one here.

Let d, e, and f be the roots of the cubic $x^3 - 6x^2 + 11x + a - 6 = 0$. Then we have

$$x^3 - 6x^2 + 11x + a - 6$$
$$= (x - d)(x - e)(x - f)$$
$$= x^3 - (d + e + f)x^2 + (de + ef + fd)x + def.$$

Hence,

$$d + e + f = 6,$$
$$de + ef + fd = 11,$$
$$def = 6 - a.$$

Notice that

$$(d + e + f)^2 = d^2 + e^2 + f^2 + 2(de + ef + fd)$$
$$\text{or} \quad 36 = d^2 + e^2 + f^2 + 22,$$
$$\therefore \quad 14 = d^2 + e^2 + f^2.$$

This certainly implies that $-3 \le d, e, f \le 3$.

Since d, e, and f are distinct and since $d + e + f = 6$, the only solution is $\{d, e, f\} = \{1, 2, 3\}$, which implies that $6 - a = def = 1(2)(3) = 6$, so that $a = 0$.

(227) *Problem.* Besides the trivial pairs $(2, 2)$ and $(2, 2)$, find two pairs of positive integers such that the sum of either pair is the product of the other pair.

Solution. Let the pairs be (x, y) and (z, w). We must have both

$$x + y = zw \quad \text{and} \quad z + w = xy.$$

Rearranging each equation and adding we obtain

$$zw - x - y + xy - z - w = 0,$$
$$(zw - z - w + 1) + (xy - x - y + 1) = 2,$$
$$(z - 1)(w - 1) + (x - 1)(y - 1) = 2.$$

Since the variables are all positive integers, both terms are non-negative. If both are *positive*, then each term equals 1, and we have the trivial solution $x = y = z = w = 2$. On the other hand, if one of the terms is 0, then the other is 2. But 2 can arise as a product only if the factors are 1 and 2, which corresponds to the integers 2 and 3. Since this pair has sum 5, then the other pair must be two integer factors of 5, namely 1 and 5, yielding the only other solution

$$(2, 3) \quad \text{and} \quad (1, 5).$$

(228) *Problem.* Let A be any subset of $N = \{1, 2, 3, \ldots, n\}$, and let the members of A be arranged in decreasing order of magnitude. Now form a sum S by alternately adding and subtracting successive members in this arrangement. For example, the subset $A = \{11, 6, 17, 1, 9, 18, 13\}$, for $n = 20$, yields the sum

$$S = 18 - 17 + 13 - 11 + 9 - 6 + 1 = 7.$$

What is the sum of all such alternating sums that are generated by the entire set $N = \{1, 2, 3, \ldots, n\}$?

Solution. First we need to recognize that the number of subsets of a set of n elements is 2^n (counting the empty set).

Thus, there are 2^n alternating sums S to be dealt with: half of them (namely 2^{n-1}) do NOT involve the integer n itself, and the remaining half (also 2^{n-1}) DO involve the integer n. Clearly, a sum that involves n must start off with n in the leading position. In fact, corresponding to a sum S which contains n, for example

$$S = n - a + b - c + \cdots,$$

there is the sum $S' = a - b + c - \cdots$, containing all the same integers except n, with all the signs in S' reversed. Thus,

$$S + S' = n.$$

Since the sums in each of the 2^{n-1} pairs (S, S') add to n, the grand total of all the alternating sums must be just $n \cdot 2^{n-1}$.

(229) *Problem.* A number of unit cubes are put together to make a larger cube and then some of the faces of the larger cube are painted. After the paint dries, the larger cube is taken apart. It is found that n small cubes have no paint on any face. If $n = p^2 \cdot q \cdot r$, where p, q, r are distinct prime numbers, how large was the original cube? (Provide all possible solutions.)

Solution. This problem was not intended to be as open-ended as it turned out to be. When Jim Totten originally proposed the problem, he had missed the scenario which leads to all of the infinite sets of solutions. He thanks all of those who brought the possibility of infinitely many solutions to his attention.

Since only the outer faces of the original cube could have been painted initially, the number n represents the number of unit cubes in the remaining rectangular solid after the painted sides were removed, and its dimensions can differ by at most 2 units.

Therefore, the dimensions of the rectangular solid must be one of: $p^2 \times q \times r$, $p \times p \times qr$, or $p \times pq \times r$. In the latter case we must have $pq - p \leq 2$, which forces $p = q = 2$, which is impossible since the primes are distinct.

Now let us suppose that the dimensions are $p^2 \times q \times r$. Without loss of generality, we may assume that $q < r$. Since they are both prime, they are either the primes 2 and 3 or both are odd.

The first case would force $p = 2$ as well, which is impossible. Therefore, we have $q + 1 = p^2 = r - 1$. But $q = p^2 - 1 = (p-1)(p+1)$ and q prime implies that $p - 1 = 1$, or $p = 2$.

Thus, the only solution in this case is $p = 2$, $q = 3$ and $r = 5$. In this case the original cube was a $5 \times 5 \times 5$ cube with 3 faces (not all mutually adjacent) painted.

The last possibility is that the dimensions are $p \times p \times qr$. Without loss of generality we will (again) assume that $q < r$. Since the 3 dimensions can differ by at most 2 units, we have $|qr - p| \leq 2$.

Let us consider first the possibility that $qr - p = \pm 1$, or $qr = p \pm 1$. Clearly, we must have $p \neq 2$ (otherwise, the right hand side is 1 or 3, neither of which is product of two primes).

Thus, p is odd which means that qr is even, which implies that $q = 2$ since $q < r$ and both are prime. Therefore, $p = 2r \mp 1$. This leads to an infinite set of solutions, the first four of which (ordered on p) are: $(p, q, r) = (5, 2, 3)$, $(7, 2, 3)$, $(11, 2, 5)$, $(13, 2, 7)$, which are associated with the (unpainted) rectangular solids $5 \times 5 \times 6$, $7 \times 7 \times 6$, $11 \times 11 \times 10$, $13 \times 13 \times 14$, respectively.

Now let us consider $qr - p = \pm 2$, or $qr = p \pm 2$. Clearly, $p \neq 2$ which implies that all of p, q, r are odd primes. This again leads to a possibly infinite set of solutions, the first four of which (ordered first on p, and then on q) are $(p, q, r) = (13, 3, 5)$, $(17, 3, 5)$, $(19, 3, 7)$, $(23, 3, 7)$, which are associated with the (unpainted) rectangular solids $13 \times 13 \times 15$, $17 \times 17 \times 15$, $19 \times 19 \times 21$, $23 \times 23 \times 21$, respectively.

(230) *Problem.* Determine all integers n such that

$$n^4 - 4n^3 + 22n^2 - 36n + 18$$

is a perfect square.

Solution. We first notice that

$$n^4 - 4n^3 + 22n^2 - 36n + 18 = (n^2 - 2n + 9)^2 - 63.$$

Suppose this the square of an integer k. We can certainly assume that $k > 0$. Then we have

$$(n^2 - 2n + 9)^2 - 63 = k^2,$$
$$(n^2 - 2n + 9)^2 - k^2 = 63,$$
$$(n^2 - 2n + 9 - k)(n^2 - 2n + 9 + k) = 63,$$

where the last factorization uses the difference of squares. Since it is clear that $n^2 - 2n + 9 - k < n^2 - 2n + 9 + k$ (recall that $k > 0$), if we let $63 = a \cdot b$ where $a < b$, then we have

$$n^2 - 2n + 9 - k = a \qquad \text{and} \qquad n^2 - 2n + 9 + k = b.$$

By subtracting these two equations we get $2k = b - a$. Now the only integer factorizations of 63 are $1 \cdot 63$, $3 \cdot 21$, and $7 \cdot 9$. Therefore, we have only 3 possibilities: $2k = 62$, $2k = 18$, and $2k = 2$. If $2k = 62$, we have $k = 31$ and $n^2 - 2n + 9 - 31 = 1$, that is, $n^2 - 2n - 23 = 0$, which clearly has no integer solutions. If $2k = 18$, we have $k = 9$ and $n^2 - 2n + 9 - 9 = 3$, that is, $n^2 - 2n - 3 = 0$, which has solutions $n = 3$ and $n = -1$. If $2k = 2$, we have $k = 1$ and $n^2 - 2n + 9 - 1 = 7$, that is, $n^2 - 2n + 1 = 0$, which has the unique solution $n = 1$. Thus, the only integers n for which $n^4 - 4n^3 + 22n^2 - 36n + 18$ is a perfect square are $n = -1$, 1, and 3.

(231) *Problem.* How many years in the 21st century have the property that when the year is divided by each of 2, 3, 5, and 7, the remainder is always equal to 1?

Solution. Suppose that year n has the desired property. Then $n - 1$ is evenly divisible by each of 2, 3, 5, and 7. Since 2, 3, 5, and 7 are distinct prime numbers, we see that $n - 1$ is evenly divisible by their product, namely $2 \times 3 \times 5 \times 7 = 210$. Therefore, n is 1 greater than some multiple of 210. But, it is clear that 2101 satisfies this condition, and the previous integer which satisfies this condition is $2101 - 210 = 1891$. Since the years 1891 and 2101 both lie outside of the 21st century, there are no years in the 21st century which leave a remainder of 1 when divided by each of 2, 3, 5, and 7.

(232) *Problem.* Find all triples of integers satisfying the inequality:

$$x^2 + y^2 + z^2 + 3 < xy + 3y + 2z.$$

Solution. By completing the squares, the inequality can be rewritten:

$$x^2 - xy + y^2 - 3y + z^2 - 2z + 3 < 0,$$

$$\left(x^2 - xy + \frac{y^2}{4}\right) + \frac{3}{4}(y^2 - 4y) + (z^2 - 2z) + 3 < 0,$$

$$\left(x - \frac{y}{2}\right)^2 + \frac{3}{4}(y^2 - 4y + 4) + (z^2 - 2z + 1) - 1 < 0,$$

$$\left(x - \frac{y}{2}\right)^2 + \frac{3}{4}(y - 2)^2 + (z - 1)^2 < 1. \tag{1}$$

Since the terms on the left hand side are all non-negative, we see that the terms

$$x - \frac{y}{2}, \qquad \frac{\sqrt{3}(y - 2)}{2}, \qquad \text{and} \qquad z - 1$$

all lie between -1 and 1. That is,

$$0 < z < 2, \qquad 2 - \frac{2}{\sqrt{3}} < y < 2 + \frac{2}{\sqrt{3}}, \qquad \text{and} \qquad -2 < 2x - y < 2,$$

from which it follows that $z = 1$, $y = 1$, 2, or 3. With $z = 1$, the inequality (1) simplifies to

$$\left(x - \frac{y}{2}\right)^2 + \frac{3}{4}(y - 2)^2 < 1.$$

If $y = 1$ or 3, then we further obtain:

$$\left(x - \frac{y}{2}\right)^2 + \frac{3}{4} < 1,$$

which implies that $|x - (y/2)| < 1/2$, or $|2x - y| < 1$. Since x and y are integers, this means that $2x = y$, which is impossible, since y is odd. Therefore, we have $y = 2$, which forces $x = 1$, and our only solution is $(x, y, z) = (1, 2, 1)$.

(233) *Problem.* Show that, regardless of what integers are substituted for x and y, the expression

$$x^5 - x^4 y - 13x^3 y^2 + 13x^2 y^3 + 36xy^4 - 36y^5$$

is never equal to pq, where p and q are prime numbers.
(A prime number is an integer greater than 1 which has no integer divisors except itself and 1, and their negatives.)
Solution. The given expression may be factored as follows:

$$\begin{aligned}
N &= x^5 - x^4 y - 13x^3 y^2 + 13x^2 y^3 + 36xy^4 - 36y^5 \\
&= x^4(x - y) - 13x^2 y^2(x - y) + 36y^4(x - y) \\
&= (x - y)(x^4 - 13x^2 y^2 + 36y^4) \\
&= (x - y)(x^2 - 4y^2)(x^2 - 9y^2) \\
&= (x - y)(x + 2y)(x - 2y)(x + 3y)(x - 3y).
\end{aligned}$$

If $y = 0$, then $N = x^5$, which is never equal to the product of two primes. On the other hand, if $y \neq 0$, then the five factors of N are all distinct.

However, any expression of pq (with p and q prime) as a product of distinct integers contains at most four factors, specifically as $(\pm 1)(\pm 1)(p)(q)$, $(\pm 1)(\pm 1)(-p)(-q)$, or $(1)(-1)(\pm p)(\mp q)$. Therefore, for any choice of integers x and y, N is never equal to a product of two primes.

(234) *Problem.*   Show that all numbers of the form:

$$10001, 100010001, 1000100010001, \ldots$$

where there are three zeroes between the ones, are composite numbers. Recall that a composite number is an integer which can be expressed as a product of two integers, each larger than 1.

(Note: all the above numbers are expressed in decimal notation.)

Solution.   All the numbers we are interested in can be expressed in the form:

$$1 + 10^4 + 10^8 + \cdots + 10^{4k},$$

for some positive integer k. Along with these numbers we shall also investigate integers of the form

$$1 + 10^2 + 10^4 + 10^6 + \cdots + 10^{2k}$$

for all positive integers k. It can be easily verified that

$$10^{4k+4} - 1 = (10^4 - 1)(1 + 10^4 + 10*8 + \cdots + 10^{4k}),$$
$$10^{2k_2} - 1 = (10^2 - 1)(1 + 10^2 + 10^4 + \cdots + 10^{2k}).$$

It is also clear that

$$10^{4k+4} - 1 = (10^{2k+2} - 1)(10^{2k+2} + 1).$$

Comparison of these equations yields:
$$(10^4 - 1)(1 + 10^4 + 10^8 + \cdots + 10^{4k}) = 10^{4k+4} - 1$$
$$= (10^2 - 1)(1 + 10^2 + 10^4 + \cdots + 10^{2k})(10^{2k+2} + 1).$$

Since $(10^4 - 1)/(10^2 - 1) = 10^2 + 1 = 101$, we have

$$(1+10^4+10^8+\cdots+10^{4k}) \times 101 = (1+10^2+10^4+\cdots+10^{4k})(10^{2k+2}+1).$$

Since 101 is a prime number, at least one of the two factors on the right hand side is divisible by 101. If $k > 1$, then whichever of these two numbers is divisible by 101, the quotient will exceed 1; hence, $1 + 10^4 + 10^8 + \cdots + 10^{4k}$ is, for $k > 1$, expressible as a product of two factors, each greater than 1. If $k = 1$, we have that the number $10^4 + 1 = 10001$, which is composite ($10001 = 73 \times 137$).

(235) *Problem.*   Find the sum of all four-digit numbers which contain only the digits 1, 2, 3, 4, 5, each occurring at most once.

Solution. Since there are 5 ways to select the first digit, 4 ways to select the second, etc., there are a total 120 such numbers. Furthermore, each digit will appear in each of the four positions in exactly one-fifth (or 24) of the these numbers.

Thus, each column sum will be $24 \times (1+2+3+4+5) = 24 \times 15 = 360$. Therefore, the total of all such numbers will be

$$360 \times (1000 + 100 + 10 + 1) = 360 \times 1111 = 399{,}960.$$

Note: $33,333$ is the mean, since in each place value, the mean is the digit 3. The sum must be $120 \times 33,333 = 399,960$.

(236) *Problem.* The increasing sequence 1, 5, 6, 25, 26, 30, 31, 125, 126, ... consists of positive integers that can be formed by adding distinct powers of 5. What is the 75th integer in the sequence?

Solution. If we were to write these numbers in base 5 notation, each of them would be a number with only the digits 0 and 1. Furthermore, any base 5 number made up of only 0s and 1s appears (eventually) in the sequence. Since the only digits appearing are 0 and 1, each such base 5 number could also be interpreted as a binary (base 2) number. Conversely, every binary number could be interpreted as a base 5 number with only the digits 0 and 1 appearing in its representation. All of the above means that the 75th number in the given sequence has the same digit pattern as the number 75 written in binary. Since 75 in binary is 1001011 (that is, $2^6 + 2^3 + 2^1 + 2^0$), the 75th number in the given sequence is

$$5^6 + 5^3 + 5^1 + 5^0 = 15625 + 125 + 5 + 1 = 15756.$$

(237) *Problem.* Find all positive integers a, b, c, d, and e which satisfy

$$a! = b! + c! + d! + e!.$$

Solution. Clearly, a must be the largest integer. Then, without loss of generality we may suppose that $a > b \geq c \geq d \geq e \geq 1$. It follows from the given equation that $a! \geq 4e! \geq 4$, implying that $a \geq 3$.

Since $a > b$, we have $a! \geq a(b!)$. The given equation yields $a! \leq 4b!$, which thus implies that $a \leq 4$. Therefore, we have either $a = 3$ or $a = 4$.

When $a = 3$, we have $b! + c! + d! + e! = 6$. It is easy to see that a, b, c, and d must be some arrangement of the numbers 2, 2, 1, and 1, after we lift the restriction on the order of b, c, d, and e.

When $a = 4$, we have $b! + c! + d! + e! = 24$. All of b, c, d, and e must be strictly less than 4; on the other hand, they must be at least 3 since $4 \cdot 3! = 24$. Therefore, $a = 4$ and $b = c = d = e = 3$ is also a solution.

(238) *Problem.*   For how many positive integers x does there exist a positive integer y with

$$\frac{xy}{x + y} = 100\,?$$

Solution.   Clearing the fraction yields $xy = 100x + 100y$, or $xy - 100x - 100y = 0$. The latter can be "completed" to yield:

$$xy - 100x - 100y + 100^2 = 100^2$$

$$(x - 100)(y - 100) = 10000.$$

This means that $x - 100$ is divisor of 10000. The positive divisors of 10000 can be listed as:

1, 2, 4, 5, 8, 10, 16, 20, 25, 40, 50, 80, 100, 125, 200,

250, 400, 500, 625, 1000, 1250, 2000, 2500, 5000, 10000.

Each of these 25 values yield a solution with x positive. On the other hand, if one of $x - 100$, $y - 100$ is negative, then so is the other since their product is positive. However, since we require that x and y are positive, we see that $x - 100 \geq -99$ and $y - 100 \geq -99$, whose product will be at most 99^2, which is less than 10000. Therefore, there are only the 25 positive divisors listed above which need to be considered. The values of x can then be calculated as:

101, 102, 104, 105, 108, 110, 116, 120,

125, 140, 150, 180, 200, 225, 300, 350, 500,

600, 725, 1100, 1350, 2100, 2600, 5100, 10100.

(239) *Problem.*   Elizabeth has a secret hiding place where she presently has 35 coins, 38 baseball cards, and 39 pieces of candy. Her younger brother, John, has his own secret hiding place, but that does not prevent him from taking things from Elizabeth's (her place is not as secret as she thinks).

John always takes two items at a time, each of a different kind so that the number will not decrease too quickly, and he always adds one item of the third kind.

One day, some time later, Elizabeth was shocked to find that all the items in her hiding place were of the same kind. Was it all candy?

Solution. Note that each time John makes changes to Elizabeth's treasures, he alters the parity of the numbers of each item (parity means the even-odd nature of the number).

Thus, the parity of the coins and candy will always be the same, and the number of baseball cards will be of opposite parity. At the time when she discovers only one type of item in her hiding place, the number of items of the other two types have both gone to 0 (an even number).

This means that the coins and the candy have both disappeared, and the only item remaining is baseball cards, and not candy.

(240) *Problem.* Find all 3-digit integers n (no leading zeroes) such that the sum of the squares of the digits of n is exactly one-half of n.

Solution. Let the digits of such an integer n be a, b, and c; that is, the number n is abc in base 10 positional notation.

Thus, $n = 100a + 10b + c$. The given condition then implies that $100a + 10b + c = 2a^2 + 2b^2 + 2c^2$. Hence, c must be even, so that $c = 2k$ and $c \le 8$.

Note that $a^2 + b^2 + c^2 \le 3(81) = 243$, and therefore, $n \le 2(243) = 486$. Therefore, $a \le 4$ and $a^2 + b^2 + c^2 \le 16 + 81 + 64 = 161$. Now we have $n \le 2(161) = 322$.

We consider the three cases where $a = 3$, 2, and 1, respectively:

If $a = 3$, we have $\frac{300}{2} \le 9 + b^2 + c^2 \le \frac{398}{2}$.

Since c is even, the only values of (a, b, c) are $(3, 9, 8)$.

From the inequality, we can see that $a^2 + b^2 + c^2 \ne \frac{398}{2}$.

If $a = 2$, we have $\frac{200}{2} \le 4 + b^2 + c^2 \le \frac{198}{2}$.

Since c is even, the possible values of (a, b, c) are here:

$(2, 9, 8)$; $(2, 8, 8)$; $(2, 7, 8)$; $(2, 6, 8)$; $(2, 9, 6)$; $(2, 8, 6)$ and $(2, 9, 4)$.

Checking the sums of the squares of these digits, we find

$$2(2^2 + 9^2 + 8^2) = 298,$$

and no other values give the desired relationship.

(241) *Problem.* Find all natural numbers whose square (in base 10) is represented by odd digits only.

Solution. We only need to consider odd natural numbers. Clearly, 1 and 3 both satisfy the given condition. Conversely, let n be such a natural number, and let y be its units digit (y is obviously one of 1, 3, 5, 7, or 9). Then $n = 10x + y$ for some integer x, and we have

$$n^2 = (10x + y)^2 = 100x^2 + 20xy + y^2.$$

Clearly, the units digit of n^2 is the same as the units digit of y^2, and the tens digit of n^2 is an even number plus whatever "carry" there is from the square of y; that is, plus the tens digit of y^2. But for all the above values of y, the tens digit of y^2 is even. Thus, the tens digit of n is even, which is a contradiction unless $x = 0$ and y^2 has no "carry". That is, n must be either 1 or 3.

(242) *Problem.* Find all sets of five positive integers whose sum equals their product. Prove that you have obtained all solutions.

Solution. Let x_1, x_2, x_3, x_4, and x_5 be the positive integers whose sum and product are equal. Suppose that $x_1 \leq x_2 \leq x_3 \leq x_4 \leq x_5$. Then $x_1 x_2 x_3 x_4 x_5 = x_1 + x_2 + x_3 + x_4 + x_5 \leq 5x_5$. Thus, $x_1 x_2 x_3 x_4 \leq 5$. This leaves us with 5 cases.

Case 1: $x_1 x_2 x_3 x_4 = 1$.
This implies that $x_1 = x_2 = x_3 = x_4 = 1$, which means that $x_5 = 4 + x_5$, a contradiction. Thus, this case is impossible.

Case 2: $x_1 x_2 x_3 x_4 = 2$.
This implies that $x_4 = 2$ and $x_1 = x_2 = x_3 = 1$. Hence, $2x_5 = 5 + x_5$, which yields $x_5 = 5$, and the solution $(1, 1, 1, 2, 5)$.

Case 3: $x_1 x_2 x_3 x_4 = 3$.
As in case 2, this implies that $x_4 = 3$ and $x_1 = x_2 = x_3 = 1$. Thus, $3x_5 = 6 + x_5$, which yields $x_5 = 3$, and the solution $(1, 1, 1, 3, 3)$.

Case 4: $x_1 x_2 x_3 x_4 = 4$.
This implies that we have either (a) $x_4 = 4$ and $x_1 = x_2 = x_3 = 1$, or (b) $x_3 = x_4 = 2$ and $x_1 = x_2 = 1$. In the former case, we get $4x_5 = 7 + x_5$, which has no solution in integers; in the latter case, we get $4x_5 = 6 + x_5$, which yields $x_5 = 2$. Thus, we have the solution $(1, 1, 2, 2, 2)$.

Case 5: $x_1 x_2 x_3 x_4 = 5$.
As in cases 2 and 3 above, we get $x_4 = 5$ and $x_1 = x_2 = x_3 = 1$. Thus, $5x_5 = 8 + x_5$, which yields $x_5 = 2$. Since we are assuming that $x_4 \leq x_5$, this produces no solution (alternatively, we could produce the solution $(1, 1, 1, 5, 2)$, which has already been accounted for in case 2).

In conclusion, we have exactly three solutions: $(1, 1, 1, 2, 5)$, $(1, 1, 1, 3, 3)$, and $(1, 1, 2, 2, 2)$.

The first solution above can be generalized to the more general problem of finding a set of n positive integers whose sum equals its product. A set of n numbers which will always satisfy this is the set consisting of the integer n, the integer 2, and $n - 2$ copies of the integer 1, giving both a sum and product of $2n$.

(243) *Problem.* Divide the <u>positive</u> integers into two disjoint subsets A and B such that a positive integer belongs to the subset A if and only if it is the sum of two different elements of A or the sum of two different elements of B.

Determine the set A.

Solution. Since 1 and 2 cannot be expressed as the sum of two different positive integers, they must belong to the set B. This forces the number 3 into the set A. The number 4 must then also belong to B, which means that 5 and 6 belong to A.

The number 7 can only be expressed in three ways as the sum of positive integers, namely $1 + 6$, $2 + 5$, and $3 + 4$, each of which has a number from A and one from B. Consequently, 7 belongs to the subset B.

We see that 8 and 9 belong to A (in more than one way!). As was the case with 7, the number 10 can only be expressed as a sum of positive integers with one summand in A and the other in B. Thus, 10 belongs to B.

Now, we can check that 11, 12, and 13 all belong to A. Since 3 also belongs to A, by simply adding multiples of 3 to these, and then to the resulting sums, etc., we are able to successively show that all the remaining positive integers belong to A.

In summary, we see that A consists of all positive integers with the exception of 1, 2, 4, 7, and 10.

(244) *Problem.* To simply the expression

$$\frac{37^3 + 13^3}{37^3 + 24^3},$$

a student has incorrectly "cancelled the exponents" as

$$\frac{37^3 + 13^3}{37^3 + 24^3} = \frac{37 + 13}{37 + 24} = \frac{50}{61},$$

and obtained the correct answer.

Find necessary and sufficient conditions for the positive integer triple (A, B, C) to satisfy

$$\frac{A^3 + B^3}{A^3 + C^3} = \frac{A + B}{A + C}.$$

Solution. For positive real numbers A, B, and C, we have

$$\frac{A^3 + B^3}{A^3 + C^3} = \frac{A + B}{A + C}$$
$$(A + C)(A^3 + B^3) = (A + B)(A^3 + C^3)$$
$$(A + C)(A + B)(A^2 - AB + B^2) = (A + B)(A + C)(A^2 - AC + C^2)$$
$$A^2 - AB + B^2 = A^2 - AC + C^2$$
$$B^2 - C^2 - AB + AC = 0$$
$$(B - C)(B + C - A) = 0.$$

Therefore, the condition we need is that either $B = C$ or $A = B+C$. Note that there is no need to specify that A, B, and C are integers. They may be any positive real numbers.

(245) *Problem.* How many solutions in the real numbers does the following system have:

$$x + y = 2,$$
$$xy - z^2 = 1\,?$$

Solution. From the first of the given equations we get

$$x = 2 - y.$$

Upon substitution into the second, we obtain

$$2y - y^2 - z^2 = 1,$$

or

$$z^2 + y^2 - 2y + 1 = 0,$$

or

$$z^2 + (y - 1)^2 = 0.$$

Each of the two terms of the last equation is non-negative, hence both must vanish. Thus, $z = 0$ and $y = 1$, which implies that $x = 1$. Therefore, the system has precisely one real solution.

(246) *Problem.* Prove that if n is a positive integer, then

$$\frac{n^2 + 3n + 1}{n^2 + 4n + 3}$$

is an irreducible fraction.

Solution. Suppose instead that the fraction can be reduced. This means that the numerator and denominator have a common integer factor larger than 1. Since it is larger than 1 it has a prime factorization. Let p be any prime factor of this common divisor. Then p divides evenly into both the numerator and denominator, and therefore, p must also divide evenly into their difference, namely $n + 2$.

On the other hand the denominator factors into $(n + 1)(n + 3)$. Since p evenly divides that denominator and p is prime it must divide either $n + 1$ or $n + 3$ (or both).

In either case, since p evenly divides $n + 2$ also, it must divide the difference between $n + 2$ and either $n + 1$ or $n + 3$, which is ± 1. This is impossible since p is prime. Therefore, the fraction must be irreducible.

(247) *Problem.* Positive integers m and n are given. When m is divided by n the quotient is a_1 and the remainder is r_1. When $m + r_1$ is divided by n the quotient is a_2 and the remainder is r_2; dividing $m + r_2$ by n yields a_3, r_3, similarly; and we continue until we obtain a_n, r_n. Find the value of

$$a_1 + a_2 + \cdots + a_n.$$

Solution. Write

$$m = a_1 n + r_1,$$
$$m + r_1 = a_2 n + r_2,$$
$$m + r_2 = a_3 n + r_3,$$
$$\vdots$$
$$m + r_{n-1} = a_n n + r_n.$$

Now sum these equations to get

$$mn + (0 + r_1 + 2_2 + \cdots + r_{n-1}) = n(a_1 + a_2 + \cdots + a_n) + (r_1 + r_2 + \cdots + r_n)$$

which reduces to

$$m = (a_1 + a + 2 + \cdots + a_n) + \frac{r_n}{n}.$$

Since m and each a_i is an integer, it follows that r_n/n is an integer also. However, we also have that $0 \le r_i < n$ for every i; this implies that $r_n = 0$, from which it follows that

$$a_1 + a_2 + \cdots + a_n = m.$$

(248) *Problem.* Two similar triangles with integral sides have two of their sides the same length. The third sides differ by 387. What are the lengths of the sides?

Solution. Let the sides of the smaller triangle be a, b, and c with $a < b < c$. Then the sides of the larger triangle must be b, c, and d with $a/b = b/c = c/d$. Let the common value be denoted by m/n in the lowest terms. Then

$$\frac{a}{d} = \frac{a}{b} \times \frac{b}{c} \times \frac{c}{d} = \frac{m^3}{n^3}.$$

Hence, $a = km^3$ and $d = kn^3$ for some integer k. Now

$$3^2 \times 43 = 387 = d - a = k(n^3 - m^3) = k(n - m)(n^2 + nm + m^2),$$

where both 3 and 43 are prime.

Since $(n - m)^2 = n^2 - 2nm + m^2 < n^2 + nm + m^2$, we see that $n - m$ must be either 1 or 3. Suppose first that $n - m = 1$ or $n = m + 1$. This implies that

$$387 = k(n^2 + nm + m^2) = k(m^2 + 2m + 1 + m^2 + m + m^2)$$
$$= k(3m^2 + 3m + 1).$$

Since the factor $3m^2 + 3m + 1$ leaves a remainder of 1 on division by 3, and yet it must divide $387 = 3^2 \times 43$, we must have $3m^2 + 3m + 1 = 43$, but this does not have an integer solution for m. Therefore, we conclude that $n - m = 3$, or $n = m + 3$. This implies that

$$129 = k(m^2 + 6m + 9 + m^2 + 3m + m^2) = k(3m^2 + 9m + 9),$$

which is equivalent to $k(m^2 + 3m + 3) = 43$. Since $m^2 + 3m + 3 > 3$, we must have $k = 1$ and $m^2 + 3m + 3 = 43$, from which it follows that $m = 5$ and $n = 8$ (since the negative root $m = -8$ is extraneous). Then, $a = 125$, $b = 200$, $c = 320$, and $d = 512$.

(249) *Problem.* Find all five-digit numbers such that when multiplied by 9, the product is given by writing the five digits of the number in reverse order.

Solution. Let $abcde$ be the decimal digit representation of the number(s) n that we seek. Then $9 \times (abcde)_{10} = (edcba)_{10}$, which becomes:

$$9(10000a + 1000b + 100c + 10d + e)$$

$$= 10000e + 1000d + 100c + 10b + a. \qquad (1)$$

Normally, we assume that a five-digit number actually has 5 digits (that is, it does not have any leading zeroes).

Let us handle that case first. It is easy to see that $a \leq 1$, for otherwise $9n$ would have greater than 5 digits.

Thus, $a = 1$, which implies that $e = 9$ since e is the leading digit of the product. Then (1) becomes:

$$90000 + 9000b + 900c + 90d + 81$$
$$= 90000 + 1000d + 100c + 10b + 1,$$
$$8990b + 800c + 80 = 910d,$$
$$899b + 80c + 8 = 91d.$$

All of the terms in the above equation are positive, and the largest we can have for the right hand side occurs when $d = 9$, namely 819.

This implies that b must be 0. Our equation then becomes:

$$80c - 91d = 8.$$

Checking the various digits for c and d, we see that there is only one solution among decimal digits for the equation, namely $c = 9$ and $d = 8$. Thus, $n = 10989$.

Now let us consider the case where n has leading zeroes, that is, $a = 0$. In this case, (1) becomes

$$9000b + 900c + 90d + 9e = 10000e + 1000d + 1000c + 10b.$$

Since the right hand side is a multiple of 10 we must have $e = 0$. Then the above becomes

$$9000b + 900c + 90d = 1000d + 100c + 10b,$$
$$8990b + 800c = 910d,$$
$$8996 + 80c = 91d.$$

We can then argue as above that $b = 0$ since the largest value we can have on the right hand side is 819. The whole equation then reduces to

$$80c = 91d.$$

This forces 91 to be a divisor of c, which can only occur if $c = 0$ (since c is a single decimal digit). It then clearly follows that $d = 0$ and the only such n would be 0.

Thus, if we allow leading zeroes the only additional solution we get is $n = 0$, which is very uninteresting!

(250) *Problem.*   A four-digit decimal number *abcd* is said to be faulty
if it has the following property: the product of the last two digits
c and d equals the two-digit number ab, while the product of the
digits $c - 1$ and $d - 1$ equals the two digit number ba. Find all
four-digit faulty numbers.

Solution.   Rewriting the given conditions as equations we have

$$10a + b = cd, \tag{1}$$

$$10b + a = (c - 1)(d - 1) = cd - c - d + 1. \tag{2}$$

Since we started with a four-digit number we have $a \geq 1$. Equation (2) then implies that $c \geq 2$ and $d \geq 2$. Substituting (1)
into (2), and rearranging, we get

$$c + d = 1 + 9(a - b). \tag{3}$$

Since c and d are both greater than 1, we see that $a \geq b + 1$. On
the other hand, if $a \geq b + 2$, we would have $c + d \geq 19$, which is
impossible for single decimal digits.

Thus, we conclude that $a = b + 1$. Then, equations (3) and (1)
can be rewritten as $c + d = 10$ and $cd = 11b + 10$. Since c and d
are interchangeable, we need only consider the (unordered) pairs
$(2, 8)$, $(3, 7)$, $(4, 6)$, and $(5, 5)$ for (c, d). Of these only $(3, 7)$ satisfies
$cd = 11b + 10$. Thus, the only faulty numbers are 2137 and 2173.

(251) *Problem.*   Find all real number triples (x, y, z) such that when
any one of these numbers is added to the product of the other two,
the result is 2.

Solution.   The system to be solved is:

$$x + yz = 2, \tag{1}$$

$$y + zx = 2, \tag{2}$$

$$z + xy = 2. \tag{3}$$

Subtracting (2) from (1) yields

$$(x - y)(1 - z) = 0,$$

which implies that $x = y$ or $z = 1$, while subtracting (3) from (2)
yields

$$(y - z)(1 - x) = 0,$$

which implies that $y = z$ or $x = 1$.

Let us consider the 4 cases:

(i) $x - y = 0 = y - z$. This implies that $x = y = z$ and equation (1) then becomes $x^2 + x - 2 = 0$, which has solution $x = 1$ or $x = -2$. Thus, $x = y = z = 1$ or $x = y = z = -2$.

(ii) $x - y = 0 = 1 - x$. This means that $x = y = 1$. Then equation (3) becomes $z + 1 = 2$, implying $z = 1$, and we again get $x = y = z = 1$.

(iii) $1 - z = 0 = y - z$. This is similar to (ii) above.

(iv) $1 - z = 0 = 1 - x$. Thus, $x = z = 1$. Then equation (2) forces $y = 1$ and we again have $x = y = z = 1$.

Thus, the only solutions are $x = y = z = 1$ and $x = y = z = -2$.

(252) *Problem.* A pencil, eraser, and notebook together cost one dollar. A notebook costs more than two pencils, three pencils cost more than four erasers, and three erasers cost more than a notebook. How much does each item cost (assuming that each item costs an integral number of cents)?

Solution. Let P, E, and N denote the cost, in cents, of each pencil, eraser, and notebook, respectively. Note that P, E, and N are positive integers such that

$$P + E + N = 100,$$
$$N > 2P,$$
$$3P > 4E,$$
$$3E > N.$$

We have $P < \frac{N}{2} < \frac{3E}{2}$, so that

$$100 = P + E + N < \frac{3E}{2} + E + 3E = \frac{11}{2}E,$$

that is, $E > 18E$. Also $N > 2P > \frac{8E}{4}$, so that

$$100 = P + E + N > \frac{4E}{3} + E + \frac{8E}{4} = 5E;$$

that is, $E < 20$. Together the restrictions on E yield $E = 19$. Now we have

$$81 = N + P > 2P + P = 3P > 4E = 76,$$

that is, $81 > 3P > 76$, so that $P = 26$.

Finally, $N = 100 - P - E = 55$.

(253) *Problem.*   Find all integer solutions to the system of equations:

$$x^3 + y^3 + z^2 = 3 = x + y + z.$$

Solution. We first note that $(1,1,1)$ is a solution. We will now look for other solutions. In general, at least one of x, y, z must be positive; assume x is positive. And because the only three cubes of non-negative integers that sum to 3 are 1, 1, 1, we may assume that at least one of y or z is negative; assume y is negative. Because $z = 3 - (x + y)$, we have $3 = x^3 + y^3 + (3 - (x + y))^3$, which simplifies to $(x - 3)(y - 3)(x + y) = 8$. Thus, $y - 3$ divides 8 and y is negative, which means that y is -1 or -5. If $y = -1$, then $(x - 3)(x - 1) = -2$, which has no real solution. For $y = -5$, we have $(x - 3)(x - 5) = -1$, which means that $x = 4$. Then we also have $z = 3 - (x + y) = 4$. The complete solution set is therefore $(1, 1, 1)$, $(-5, 4, 4)$, $(4, -5, 4)$, and $(4, 4, -5)$.

(254) *Problem.*   Use the digits 1 through 9 once only to form a nine-digit number such that the first (leftmost) eight digits form a number divisible by 8, the first seven form a number divisible by 7, and so on. How many such numbers are there?

Solution. Suppose $a_1 a_2 a_3 a_4 a_5 a a_6 a_7 a_8 a_9$ has the desired property. Then we get immediately that $a_5 = 5$ and a_2, a_4, a_6, a_8 are even. Since that is all the even digits, the other digits are odd. By the well-known divisibility-by-3 test, $a_1 + a_2 + a_3$ and $a_4 + a_5 + a_6$ must each be divisible by 3. It follows that $a_4 a_5 a_6$ is either 258 or 654 (the other two choices, 852 and 456, are eliminated by the oddness of a_3, which would cause a failure of the divisibility-by-4 condition).

The 258 choice leads only to the possibilities:

> 147258369, 147258963, 741258369, 741258963,
> 369258147, 369258741, 963258147, and 963258741.

But the sequences containing 836, 814, or 874 are forbidden by the divisibility-by-8 condition, while the remaining two candidates fail the divisibility-by-7 condition.

Thus, the only possibility is ***654***. The divisibility-by-8 condition tells us that the form must be ***654x2*, where x is 3 or 7, and this yields the eight possibilities: 987654321, 789654321, 981654327, 189654327, 381654729, 183654729, 981654723, and 189654723. Only one of these, namely 381654729, survives the divisibility-by-7 condition, so that it is the unique solution. (Note that the solution must satisfy a divisibility-by-9 condition also, since the sum of the digits is 45.)

(255) *Problem.* A positive integer n is called an *ambitious* number if it possesses the following property: writing it down (in decimal representation) on the right of any positive integer gives a number that is divisible by n.

Find:

(a) the first 10 ambitious numbers; (b) all the ambitious numbers.

Solution. Let m be any positive integer which is relatively prime to n, that is, it has no common factors with n except 1. Since n divides the number $m \times 10^k + n$ evenly, where k is the number of digits of n, this implies that n divides $m \times 10^k$ evenly, which also implies that n divides 10^k evenly (since m and n have no common factors). We now need only consider divisors of powers of 10. By examining successive divisors of powers of 10 we obtain the first 10 ambitious numbers:

$$1, 2, 5, 10, 20, 25, 50, 100, 125, 200.$$

Or if we want to be more analytical and consider a more general solution, we first note that $n = 2^a \times 5^b$ where $a \le k$ and $b \le k$. Since n has k digits, we clearly have $10^{k-1} \le n < 10^k$. Thus,

$$2^{k-1} \times 5^{k-1} \le 2^a \times 5^b < 2^k \times 5^k.$$

If $b \le k - 2$, then

$$n = 2^a \times 5^b \le 2^a \times 5^{k-2} \le 2^k \times 10^{k-2} < 10^{k-1},$$

which is impossible. Therefore, $b \ge k - 1$.

Case (i): $b = k - 1$. Then $2^{k-1} \le 2^a$, implying that $a \ge k - 1$. This means that $a = k - 1$ or $a = k$, both of which yield ambitious numbers for $a \ge 0$.

Case (ii): $b = k$. Then $2^{k-1} \le 2^a$, implying that $a \ge k - 3$. But $b = k$ also implies that $a \ne k$. Thus, $a = k - 3$, $a = k - 2$, or $a = k - 1$, all of which yield ambitious numbers for $a \ge 0$.

From this we see that n is an ambitious number if and only if $n = p \times 10^q$, where $p = 1, 2, 5, 25$, or 125, and q is any non-negative integer.

(256) *Problem.* You are given a set of 10 positive integers. Summing nine of them in the ten possible ways we only get nine different sums: 86, 87, 88, 89, 90, 91, 93, 94, 95. Find those numbers.

Solution. Let k be the sum which occurs twice among the numbers 86, 87, 88, 89, 90, 91, 93, 94, and 95, and let Σ represent the sum of all 10 positive integers. By considering all 10 possible ways of summing nine of the ten integers as equations, and adding all 10 equations we get:

$$86 + 87 + 88 + 89 + 90 + 91 + 93 + 94 + 95 + k = 9\Sigma.$$

Thus, $9\Sigma = 813 + k$. The only permissible value of k which yields a multiple of 9 on the left hand side of the equation is $k = 87$. Thus, $\Sigma = 100$, and the 10 positive integers can be obtained by subtracting the above sums of 9 integers from the value Σ. This yields the values 14, 13, 13, 12, 11, 10, 9, 7, 6, and 5.

(257) *Problem.* The inhabitants of Rigel III use, in their arithmetic, the same operations of addition, subtraction, multiplication, and division, with the same rules of manipulation, as are used by Earth. However, instead of working with base 10, as is common on Earth, the people of Rigel III use a different base, greater than 2 and less than 10.

The diagram below is the solution to one of their long division problems, which we have copied from a school book. We have substituted letters for the notation originally used. Each letter represents a different digit, the same digit wherever it appears.

$$
\begin{array}{r}
BC \\
AB{\overline{\smash{\big)}\,CBC}} \\
\underline{AB} \\
BDC \\
\underline{BDC}
\end{array}
$$

As the answer to this puzzle, substitute the correct numbers for the letters and state the base of the arithmetic of Rigel III.

Solution. Let G be the base used on Rigel III. Then we have $B = 1$ because $AB \times B = AB$ in base G. Similarly, we have $D = 0$ since $CB - AB = BD$, that is, $C1 - A1 = 1D$, also in base G. This same equation also implies that $C = A + 1$.

The last relationship that we have is that $A1 \times C = 10C$ in base G, or $A \times C = 10$ in base G. Since $C = A + 1$, we have $A(A+1) = G$. The only value of A which yields a base G between 2 and 10 (not inclusive) is $A = 2$, which implies that $C = 3$ and $G = 6$.

(258) *Problem.* Determine all triples of consecutive binomial coefficients

$$\binom{n}{r-1}, \binom{n}{r}, \binom{n}{t+1}$$

in arithmetic progression. Here, n and r are integers with $2 \leq r+1 \leq n$.

Solution. The given condition is equivalent to finding integers n and r satisfying $2 \leq r+1 \leq n$

$$\binom{n}{r+1} - \binom{n}{r} = \binom{n}{r} - \binom{n}{r-1}.$$

By using the definition of binomial coefficients which uses factorials we get:

$$\frac{n!}{(r+1)!(n-r-1)!} - \frac{n!}{r!(n-r)!} = \frac{n!}{r!(n-r)!} - \frac{n!}{(r-1)!(n+1-r)!},$$
$$(n+1-r)(n-r) - (r+1)(n+1-r) = (r+1)(n+1-r) - (r+1)r,$$
$$(n+1-r)(n-1-2r) = (r+1)(n+1-2r),$$
$$n^2 - 1 - r - 3nr + 2r^2 = nr + n + 1 - r - 2r^2.$$
$$n^2 - 4nr + 4r^2 = n + 2,$$
$$(n-2r)^2 = n+2.$$

Thus, $n+2$ must be a perfect square. We will now show that if $n+2$ is any perfect square larger than 4, then we get not only one, but two solutions. Clearly, if $n+2 = 4$, we must have $n = 2$, whence, $n - 2r = 2$, that is, $r = 0$, which is not allowed.

Now let $n+2 = k^2$ where k is some positive integer, $k \geq 3$. Then $n - 2r = \pm k$. That is, $2r = n \pm k = k^2 - 2 \pm k$. Thus, for any $k \geq 3$, we get the two solutions:

$$n = k^2 - 2, \qquad r = \frac{k(k \pm 1)}{2} - 1.$$

The solutions for the first few perfect squares are:

$k = 3$, so that $n = 7$ and $r = 2$ or 5;

$k = 4$, so that $n = 14$ and $r = 5$ or 9;

$k = 5$, so that $n = 23$ and $r = 9$ or 14.

(259) *Problem.* Reconstruct the following exact long division problem, where the X's can represent any digit.

$$XX\,8\,XX$$
$$XXX\overline{)XXXXXXXX}$$
$$\underline{\,XXX}$$
$$XXXX$$
$$\underline{\,XXX}$$
$$XXXX$$
$$\underline{XXXX}$$

Solution. In long division, when two digits are brought down instead of one, there must be a zero in the quotient. This occurs twice, so we know at once that the quotient is $X080X$. When the divisor is multiplied by the last digit of the quotient the result is a 4-digit number. The quotient's last digit must therefore must be 9, since 8 times the divisor is a 3-digit number.

Since 8 times 125 is 1000 (a 4-digit number), the divisor must be less than 125. Since 7 times a divisor less than 125 is less than 900, we conclude that the first digit of the quotient must be greater than 7 (otherwise the difference on the third line of the division would have to be at least 3 digits). Since the product of the first digit of the quotient and the divisor is only 3 digits, the first digit must be 8. Therefore, the divisor must be 80809.

Since 123 times 80809 is a 7-digit number we conclude that the divisor must be greater than 123. This leaves the only possibility for the divisor as 124, which yields the dividend as 10020316. The rest follows:

$$80809$$
$$124\overline{)10020316}$$
$$\underline{\,992}$$
$$1003$$
$$\underline{\,992}$$
$$1116$$
$$\underline{1116}$$

Chapter 7

Plane Geometry

(260) *Problem.* The radius of the inscribed circle of a triangle is 4 and the line segments into which one side is divided by the point of contact are 6 and 8.

Determine the lengths of the other two sides of the triangle.

Solution.

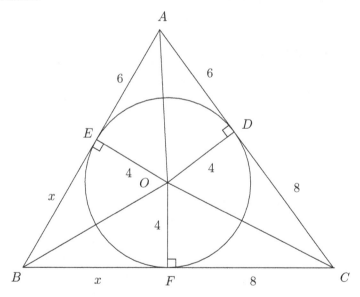

The inscribed circle (centre O) of $\triangle ABC$ is also shown above. It is given that $AD = 6$ and $DC = 8$. It follows immediately that $AE = 6$ and $CF = 8$. Let $x = BE = BF$. Now the area of $\triangle AOB$ is $2(x+6)$; that of $\triangle BOC$ is $2(x+8)$; and that of $\triangle AOC$ is $2(6+8) = 28$. Thus, the area of $\triangle ABC$ is the sum of these

215

parts, namely

$$\text{Area of } \triangle ABC \ = 2(x + 6 + x + 8 + 6 + 8) = 4(x + 14).$$

On the other hand, we know that if s is the semiperimeter of a triangle whose three sides have lengths a, b, c, then Heron's formula says

$$\text{Area} \ = \ \sqrt{s(s - a)(s - b)(s - c)}.$$

For $\triangle ABC$, we have $s = (2x + 12 + 16)/2 = x + 14$, and a, b, and c are $x + 8$, 14, and $x + 6$, respectively. Therefore,

$$\text{Area of } \triangle ABC \ = \ \sqrt{(x + 14)(6)(x)(8)}.$$

Thus, we have

$$4(x + 14) = \sqrt{(x + 14)(48x)},$$
$$16(x + 14)^2 = (x + 14)(48x),$$
$$x + 14 = 3x,$$
$$x = 7.$$

Therefore, the sides of $\triangle ABC$ have lengths 13, 14, and 15. That is, the two "other" sides have lengths 13 and 15.

(261) *Problem.* Let ABC be any triangle. Let BD (with D on AC) bisect $\angle ABC$ and let CE (with E on AB) bisect $\angle ACB$. Show that if $BD = CE$, then the triangle is isosceles.

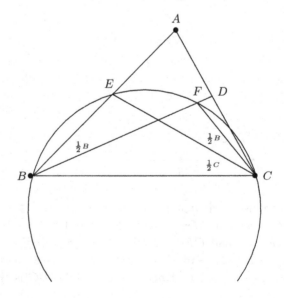

Solution. We will prove the contrapositive statement, namely if $\angle B \neq \angle C$, then $BD \neq CE$. Let ABC be any triangle with $\angle B \neq \angle C$. Without loss of generality, we may assume that $\angle B < \angle C$. We will show that $CE < BD$, where CE bisects $\angle C$ and BD bisects $\angle B$ with E on AB and D on AC.

Let F be on the line BD such that $\angle FCE = \frac{1}{2}\angle B$.

Since $\angle B < \angle C$, we see that $BF < BD$. Since $\angle EBF = \angle ECF$, we see that the points E, B, C, F lie on a circle. Note that

$$\angle B < \frac{1}{2}(\angle B + \angle C) < \frac{1}{2}(\angle A + \angle B + \angle C);$$

that is,

$$\angle CBE < \angle FCB < 90°.$$

Thus, the chord subtending $\angle CBE$ is shorter than the chord subtending $\angle FCB$. In other words, we have $CE < BF$; whence $CE < BD$, and we are done.

(262) *Problem.* A unit circle is a circle whose radius is 1. How many unit circles may be placed inside a square whose diagonal length is:

(a) $2\sqrt{2}$,   (b) $3\sqrt{2}$,   (c) $4\sqrt{2}$,   (d) 6,   (e) $2(2+\sqrt{2})$,

so that no two circles overlap? Show how you arrive at your answers in all five cases.

Solution. Note first that in order to "squeeze" as many unit circles as possible into a square, we must "push" them into the corners of the square.

(a) A square with diagonal $2\sqrt{2}$ has side length 2, and is just large enough to contain one unit circle.

(b) In order to contain two unit circles, the smallest such square would have to have the centres of these circle on its diagonal. This would force a diagonal of at least $2 + 2\sqrt{2}$ in length.

Since $2 + 2\sqrt{2} > 3\sqrt{2}$, we see that we can still fit only one unit circle in a square with diagonal length of $3\sqrt{2}$.

(c) A square with diagonal $4\sqrt{2}$ can be subdivided into four squares with diagonal $2\sqrt{2}$ and from (a) each can contain one circle. Since the space left over in the centre is not large enough to place another circle, we can have at most four unit circles in such a square.

(d) First of all, observe that $6 > 4\sqrt{2}$, so we can place at least four unit circles inside such a square. In order to place one more circle, that circle must go in the centre of the square, which means that the diagonal of the square at least $4 + 2\sqrt{2}$, which is too large. Thus, only four unit circles can fit inside a square with diagonal 6.

(e) From the argument in (d) above, we see that five unit circles can fit in a square with diagonal $2(2 + \sqrt{2}) = 4 + 2\sqrt{2}$.

(263) *Problem.* The shaded area in the figure below resembles a water wheel. If the small circle has a radius of 2 feet, and the large circle has a radius of 5 feet, calculate the area of the shaded region to the nearest hundredth of a square foot. The lines through the centre of the concentric circles form 45° angles.

Solution. Each of the shaded parts is the area between a triangle and a circular sector of the inner circle. Thus, to find the area of one shaded part, we must first find the area of the triangle. Note that the base of this triangle is 5 feet; the altitude can be calculated from the side of 2 feet and the included angle of 45° to be $\sqrt{2}$ feet. The area of one such triangle is then $\frac{5\sqrt{2}}{2}$ square feet, and that of eight such triangles is $20\sqrt{2}$ square feet. The area of the eight circular sectors is clearly the area of the smaller circle, which is 4π square feet.

Therefore, the total area is $20\sqrt{2} - 4\pi \approx 15.72$ square feet.

(264) *Problem.* A chord ST of constant length slides around a semi-circle with diameter AB. M is the mid-point of ST and P is the foot of the perpendicular from S to AB. Prove that angle SPM is constant for all positions of ST.

Solution. Enlarge the semicircle to a complete circle as shown in the diagram below.

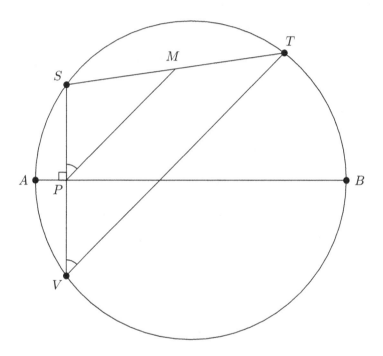

Extend the perpendicular SP through to the circumference at V. Clearly, the line segments SP and PV have the same length. Now join V to T, noting in the process that it is parallel to the line PM. Consequently, $\angle SPM = \angle SVT$. Thus, we need only show that $\angle SVT$ is constant. But $\angle SVT$ is subtended at the circumference by the chord ST, which means that it must be constant, for a well-known theorem in geometry states that the angle subtended by a chord at a point on the circumference is constant for all points on the same side of the chord.

(265) *Problem.*   In the diagram below AB and CD are of length 1 while angles ABC and CBD are 90° and 30°, respectively. Find AC.

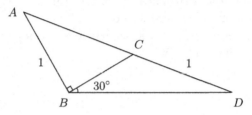

Solution.   Through the point C construct a line parallel to AB meeting the line BD at a point E.

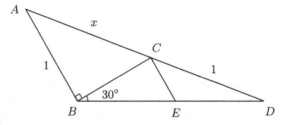

Clearly, triangles ABD and CED are similar. If we let x be the length of the line segment AC, we have

$$\frac{CD}{AD} = \frac{CE}{AB},$$

that is, $$\frac{1}{x+1} = \frac{CE}{1}.$$

Since AB and CE are parallel, we see that $\angle BCE$ is a right angle. Thus, $BC : CE = \sqrt{3}$, whence,

$$BC = \frac{\sqrt{3}}{x+1}.$$

If we now apply the Theorem of Pythagoras to triangle ABC, we get

$$AB^2 + BC^2 = AC^2,$$

or $$1 + \frac{3}{(x+1)^2} = x^2.$$

After clearing the fraction and simplifying, we obtain a polynomial which can be written in the form:

$$(x+2)(x^3 - 2) = 0.$$

From this equation we can conclude that x must be the cube root of 2 (since a value of -2 is clearly impossible!).

(266) *Problem.* Three circles of radius r each pass through the centres of the other two. What is the area of their common intersection?

Solution. If A, B, C are the three centres, then ABC is an equilateral triangle with side length r. It can be shown that the area of the triangle ABC is given by

$$\frac{\sqrt{3}}{4} r^2.$$

The area we wish to calculate is formed by three circular sectors with central angles of 60° overlapping on the triangle ABC. The area of such a circular sector is

$$\frac{\pi r^2}{6}.$$

The area in question is thus the sum of three such circular sectors less the sum of two equilateral triangles, which is

$$\frac{\left(\pi - \sqrt{3}\right) r^2}{2}.$$

(267) *Problem.* A (badly designed) table is constructed by nailing through its centre a circular disk of diameter two metres to a sphere of diameter one metre. The table tips over to bring the edge of the disk into contact with the floor. As the table rolls, the two points of contact with the floor trace out a pair of concentric circles. What are the radii of these circles?

Solution. The first point one needs to observe is that the centre of the concentric circles is located on the floor at the point where a line from the point of contact between the disk and the sphere to the centre of the sphere would meet the floor (see the figure below). Labelling the smaller radius r and the larger radius R as in the diagram, we first note that $R = r + 1$. Let h be the distance from the centre of the sphere to the centre of the concentric circles.

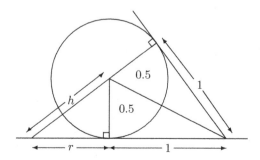

The Theorem of Pythagoras applied to the small right-angled triangle yields

$$h^2 = r^2 + (.5)^2.$$

Applying the Theorem of Pythagoras to the large right-angled triangle yields

$$(h + .5)^2 + 1^2 = R^2 = (r + 1)^2,$$

that is, $\quad h^2 + h + .25 + 1 = r^2 + 2r + 1,$

whence $\quad r^2 + (.5)^2 + h + 1.25 = r^2 + 2r + 1,$

or $\quad h = 2r - .5.$

By squaring both sides and substituting for h^2 we get

$$r^2 + (.5)^2 = 4r^2 - 2r + .25,$$

from which it follows that $r = 2/3$ and $R = 5/3$.

(268) *Problem.* In a circle of radius r construct a diameter and a chord AB parallel to the diameter (see the diagram below). Let P be any point on the diameter and let Q be the mid-point of the short arc AB. Let x be the length of the line segment PQ, let a be the length of the line segment PA and let b be the length of the line segment PB.

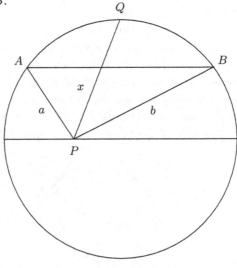

Show that

$$a^2 + b^2 = 2x^2.$$

Solution. Drop perpendiculars from A, B and Q to the diameter meeting it at the points C, D and O, respectively. Note that point O is in fact the centre of the circle and that the distances OC and OD are the same. Let d be the distance from O to D (also from O to C), let p be the distance from O to P and let e be the distance from the chord to the diameter. By applying the Theorem of Pythagoras to the right-angled triangles POQ and OBD we have two results immediately: namely,

$$r^2 + p^2 = x^2 \tag{1}$$
$$d^2 + e^2 = r^2. \tag{2}$$

By applying the Theorem of Pythagoras to the right-angled triangles PAC and PBD together we get

$$\begin{aligned}
a^2 + b^2 &= \left[(d-p)^2 + e^2\right] + \left[(d+p)^2 + e^2\right] \\
&= d^2 - 2dp + p^2 + e^2 + d^2 + 2dp + p^2 + e^2 \\
&= 2d^2 + 2p^2 + 2e^2 \\
&= 2r^2 + 2p^2 \qquad \text{from (2)} \\
&= 2x^2 \qquad\qquad \text{from (1).}
\end{aligned}$$

(269) *Problem.* Given two points A and B, find the locus of all points which are twice as far from A as from B.

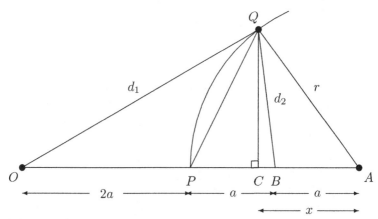

Solution. Let P be the point on the line segment AB which is twice as far from A as from B, and let it be a units from B and $2a$ units from A. Let Q be an arbitrary point on the locus. Let d_1 be the length of the line segment AQ and let d_2 be the length of the line segment BQ. Clearly $d_1 = 2d_2$. Let O be the point located on the extended line AB and a units beyond point B, and let r be the distance from O to Q. Drop a perpendicular from Q to the point C lying on the (possibly extended) line AB and let the length of this perpendicular be h. If we let x be the distance from O to C (see the diagram on the previous page), then by the Theorem of Pythagoras applied to the triangles ACQ and BCQ we get

$$(x - a)^2 + h^2 = d_2^2$$

and $$(4a - x)^2 + h^2 = d_1^2.$$

By eliminating h^2 and recalling $d_1 = 2d_2$ we get

$$d_2^2 = 5a^2 - 2ax.$$

Now by considering triangles OCQ and BCQ we have

$$x^2 + h^2 = r^2$$

or $$x^2 + d_2^2 - (x - a)^2 = r^2$$

that is, $$x^2 + 5a^2 - 2ax - (x - a)^2 = r^2,$$

whence we obtain $r^2 = 4a^2$. This clearly implies that $r = 2a$, a constant! Thus the distance r is fixed. This means that the locus is a circle of radius $2a$ centred about the point O.

(270) *Problem.* The points on the circumference of a circle are coloured in two colours, red and blue. Show that no matter how the points are coloured there are three points on the circumference which are all red or all blue and are the vertices of an isosceles triangle (that is, a triangle having at least two sides of equal length). Is it always possible to find such a triangle which is equilateral (that is, all sides equal)?

Solution. Simply construct a regular pentagon inscribed in the circle. Of the five vertices at least three must be one colour, say red. But no matter how one selects three of the vertices one must have an isosceles triangle. The answer to the second question is NO; if we colour the upper semicircle red and the lower semicircle blue and observe that an inscribed equilateral triangle cannot have all its vertices in a semicircle, then the result is obvious.

(271) *Problem.* In the figure (on the next page), the unit square $ABCD$ and the line ℓ are fixed, and the unit square $PQRS$ rotates with P and Q lying on ℓ and AB, respectively. X is the foot of the perpendicular from S to ℓ. Find the position of point Q so that

the length XY is a maximum. (A unit square is a square each of whose sides has length 1.)

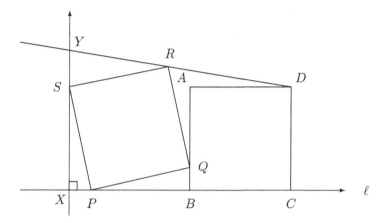

Solution. Let B be the origin and let $\theta = \angle QPB$. Then we have $\angle XSP = \theta$ and $\angle AQR = \theta$. Then XP has length $\sin\theta$ and PB has length $\cos\theta$.

Thus, the point X has coordinates $(-\sin\theta - \cos\theta, 0)$.

Similarly, point R has coordinates $(-\sin\theta, +\sin\theta + \cos\theta)$.

Since $D = (1,1)$, we can find the equation of the line RD:

$$y - 1 = \frac{1 - (\sin\theta + \cos\theta)}{1 + \sin\theta}(x - 1).$$

The point Y lies on this line. It has x-coordinate the same as the point X, namely $-\sin\theta - \cos\theta$.

If we let $f(\theta)$ be the y-coordinate of the point Y, then we have

$$\begin{aligned}
f(\theta) &= 1 + \frac{1 - (\sin\theta + \cos\theta)}{1 + \sin\theta}(-\sin\theta - \cos\theta - 1) \\
&= 1 - \frac{1 - (\sin\theta + \cos\theta)^2}{1 + \sin\theta} \\
&= 1 + \frac{2\sin\theta\cos\theta}{1 + \sin\theta}.
\end{aligned}$$

If we now differentiate with respect to θ, we have

$$f'(\theta) = \frac{(1 + \sin\theta)(2\sin\theta(-\sin\theta) + 2\cos^2\theta) - 2\sin\theta\cos\theta(\cos\theta)}{(1 + \sin\theta)^2}$$

$$= \frac{-2\sin^2\theta + 2\cos^2\theta - 2\sin^3\theta}{(1 + \sin\theta)^2}$$

$$= \frac{2(-\sin^2\theta + 1 - \sin^2\theta - \sin^3\theta)}{(1 + \sin\theta)^2}$$

$$= \frac{2(1 - 2\sin^2\theta - \sin^3\theta)}{(1 + \sin\theta^2}$$

$$= \frac{2(1 + \sin\theta)(1 - \sin\theta - \sin^2\theta)}{(1 + \sin\theta)^2}$$

$$= \frac{2(1 - \sin\theta - \sin^2\theta)}{1 + \sin\theta}.$$

This means that $f(\theta)$ has a maximum when $1 - \sin\theta - \sin^2\theta = 0$, which is a quadratic equation in $\sin\theta$.
Therefore,

$$\sin\theta = \frac{\sqrt{5} - 1}{2}.$$

Hence, the length XY is maximized when $BQ = (\sqrt{5} - 1)/2$.

(272) *Problem.*   Triangle ABC has $\angle BAC = 90°$. A circle with diameter AC meets BC at E. The tangent to the circle at E meets AB at D. Prove that EBD is an isosceles triangle. (Diagram below.)

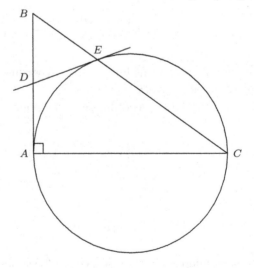

Solution. First note that AE is perpendicular to BC (since AC is a diameter of the circle). Now let P be the centre of the circle and let r be the radius. Then PE is perpendicular to DE.

Let x be the size of angle B and let y be the size of angle C. Then $x + y = 90°$. Since PEC is isosceles (2 sides of length r), we have $\angle PEC = y$.

Thus, $\angle PEA = x$, since $\angle AEC$ is a right angle. By similar arguments we can conclude that $\angle DEA = y$ and $\angle DEB = x$.

Thus, $\angle DEB = \angle DBE = x$ and triangle EBD is isosceles.

(273) *Problem.* Find the area between the circumscribed and inscribed circles of the regular 37-gon with sides of length 1.

Solution. The area is $\pi/4$, regardless of the number of sides of the n-gon. To see why, consider the following figure.

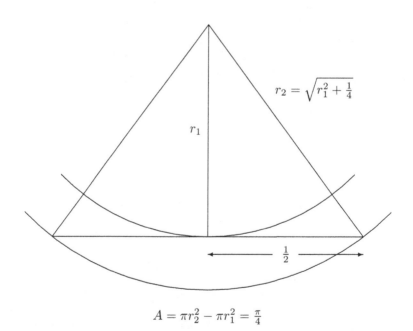

$$A = \pi r_2^2 - \pi r_1^2 = \tfrac{\pi}{4}$$

(274) *Problem.* In the diagram, OC is a radius of the larger circle and a diameter of the smaller circle which has B as centre. Prove that $CD = DE$.

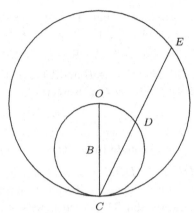

Solution. Join OD and OE. Clearly OD is perpendicular to CE, since OC is a diameter of the smaller circle. Since OC and OE are radii of the larger circle, we see that triangle OCE is isosceles. Thus, triangles OCD and OED are congruent, from which we may conclude that $CD = CE$.

(275) *Problem.*   In rectangle $ABCD$, $AD = 10$ and $CD = 15$. P is a point inside the rectangle such that $PB = 9$ and $PA = 12$. Calculate the length of PD.

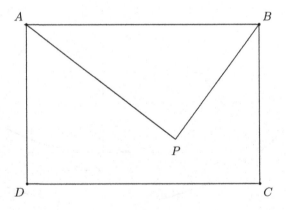

Solution. Note first that $AB = CD = 15$. Next let EF be a line passing through P and parallel to BC such that E lies on AB and F lies on CD. Since $(PA)^2 + (PB)^2 = (AB)^2$, we conclude that triangle APB is right-angled. The area of triangle APB can thus be computed in two ways:

$$\frac{1}{2}(PA)(PB) = 54 \quad \text{and} \quad \frac{1}{2}(AB)(PE) = \frac{15}{2}(PE).$$

Thus, $PE = \frac{36}{5}$. Since $EF = AD = 10$, we have $PF = \frac{14}{5}$. Since triangle PAE is similar to ABP we have

$$\frac{AE}{PA} = \frac{PA}{AB},$$

from which we see that $AE = \frac{48}{5}$. If we now use the Theorem of Pythagoras in triangle PDF, we have

$$(PD)^2 = (DF)^2 + (PF)^2 = \left(\frac{48}{5}\right)^2 + \left(\frac{14}{5}\right)^2$$
$$= \frac{48^2 + 14^2}{5^2} = \frac{4}{25}(24^2 + 7^2)$$
$$= \frac{4}{25}25^2 = 100.$$

Therefore, $PD = 10$.

(276) *Problem.* Two circles C_1 and C_2 with radii r_1 and r_2, respectively, are tangent to the line p at the point P (see the diagram below). All other points of C_1 are inside the circle C_2. The line q is perpendicular to p at the point S, is tangent to C_1 at the point R, and intersects C_2 at the points M and N, with N between R and S. Prove that PR bisects $\angle MPN$.

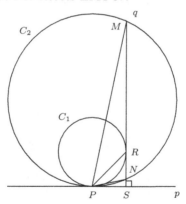

Solution. First of all note that the lines p and q are both tangent to the circle C_1. This implies that $SP = SR$. Since $\angle PSR$ is a right angle, we conclude that $\angle PRS = \angle RPS = 45°$.

Next let us consider the picture without circle C_1 and points R and S for a moment. Then $\angle PMN$ is subtended at the circumference by the chord PN. Construct a diameter of C_2 passing through P, and let M' be the other point of this diameter on the circumference.

Then $\angle PM'N = \angle PMN$, since both angles are subtended by the chord PN. But $M'P$ is perpendicular to the line p. Also $\angle PNM'$ is a right angle, from which we conclude that the angle between the line p and the line PN has the same measure as $\angle PNM'$. In the original diagram, this means that $\angle PMN = \angle NPS$.

Putting these observations together we get

$$\angle RPN = \angle RPS - \angle NPS = 45° - \angle NPS$$

$$\angle MPR = 180° - \angle PMR - \angle PRM$$

$$= 180° - \angle PMR - 135° = 45° - \angle NPS.$$

Therefore, $\angle RPN = \angle MPR$, which means that PR bisects $\angle MPS$.

(277) *Problem.* The isosceles right triangle shown below has a vertex at the centre of the square. What is the area of the common quadrilateral?

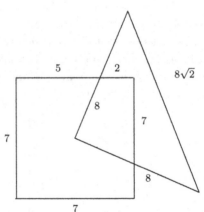

Solution. Rotating the triangle about the centre of the square does not change the common area, as what is lost in one quadrant of the square is added to an adjacent quadrant. Therefore, rotate so that the two legs of the triangle are flush with the squares diagonals (or so that they are parallel to the sides of the square). Then the common area is $\frac{1}{4}$ of the area of the square; that is, 12.25.

(278) *Problem.* Let AOB be a given sector of a circle, and C an arbitrary point on the arc AB. Let parallel lines through A and B intersect line OC at points P and Q, respectively (see the figure on the next page). Show that the area of the five-sided polygon $OAQPB$ is constant, independent of C and the parallel lines AP and BQ.

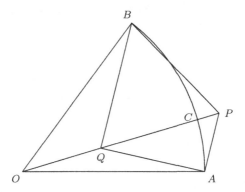

Solution. It is sufficient to show that the desired area is equal to that of $\triangle AOB$. Let M be the intersection of the lines AB and OC. We need only demonstrate that Area $\triangle AQM$ = Area $\triangle BPM$. But this is clear from the following argument: $APBQ$ is a trapezoid, and therefore Area $\triangle APQ$ = Area $\triangle APB$ since each has the same base and height; subtracting Area $\triangle APM$ from each gives the result.

(279) *Problem.* Let ABC be a right-angled triangle with a right angle at C. Drop a perpendicular from C to AB meeting AB at point G. Let d be the length of the line segment AG and let e be the length of the line segment BG. Find the value of the ratio $d : e$ in terms of angle B.

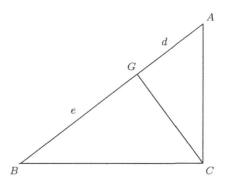

Solution. Let θ be the measure of angle B. Let a, b, and c be the lengths of the sides of triangle ABC opposite angles A, B, and C, respectively. First notice that angle ACG has measure θ. Now in triangle BCG we have $e = a \cos \theta$, while in triangle ACG we have $d = b\sin\theta$. On the other hand in triangle ABC we see that $a = c\cos\theta$ and $b = c\sin\theta$. Putting all this information together we get

$$\frac{d}{e} = \frac{b\sin\theta}{a\cos\theta} = \frac{c\sin^2\theta}{c\cos^2\theta} = \tan^2\theta.$$

(280) *Problem.*   In the figure below we are given three squares. Using only elementary geometry (not even trigonometry), prove that angle C equals the sum of angles A and B.

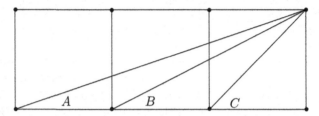

Solution.   First construct the new squares indicated in the diagram below. Angle B equals angle D because they are corresponding angles of similar right triangles. Since angles A and D add to angle C, B can be substituted for D, and it follows immediately that C is the sum of A and B.

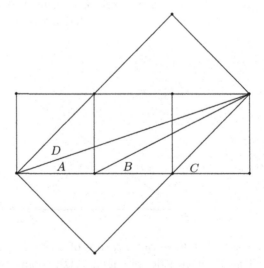

There are 54 different proofs of this known!

(281) *Problem.* AB is the diameter of a semicircle. Let P be a point on the semicircle and let Q be the point on the line AB such that PQ is perpendicular to AB.

Find the angle PAQ which gives the largest possible area for the triangle PAQ.

Solution. Let the equation of the semicircle be $x^2 + y^2 = r^2$, $y \geq 0$ where r is a positive constant.

Let us label the points $A(-r, 0)$, $B(r, 0)$, $Q(x, 0)$ and $P(x, y)$ where $x^2 + y^2 = r^2$.

The distance $PQ = y = \sqrt{r^2 - x^2}$ and the distance $AQ = r + x$.

Thus, the area of triangle APQ is given in terms of x as:

$$A(x) = (r + x)\sqrt{r^2 - x^2}.$$

Differentiating with respect to x yields:

$$A'(x) = (r + x) \cdot \frac{-x}{\sqrt{r^2 - x^2}} + \sqrt{r^2 - x^2}$$
$$= \frac{r^2 - rx - 2x^2}{\sqrt{r^2 - x^2}}.$$

To find where $A(x)$ is a maximum we set the derivative $A'(x) = 0$ and solve for x. This gives us the values of $x = -r$ or $x = \frac{1}{2}r$.

The first value for x is obviously not going to give a maximum, but rather a minimum value for $A(x)$. It can be checked that the other value for x, namely $x = \frac{1}{2}r$, does yield the maximum area.

We are asked for the value of $\angle PAQ$ which produces this maximum area.

Note that

$$\tan PAQ = \frac{y}{r + x} = \frac{\sqrt{r^2 - x^2}}{r + x} = \frac{\frac{\sqrt{3}}{2}r}{\frac{3}{2}r} = \frac{\sqrt{3}}{2} \cdot \frac{2}{3} = \frac{1}{\sqrt{3}}.$$

Since $\angle PAQ$ is obviously a first quadrant angle, we see that it must be $60°$.

(282) *Problem.* In a circle with centre O, radius OA is perpendicular to radius OB. A chord MN is drawn parallel to AB meeting OA at P and OB at Q, and the circle at M and N as shown below. If $MP = \sqrt{56}$, and $PN = 12$, determine the length of the radius of the circle. (Diagram on the next page.)

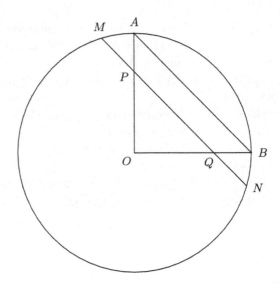

Solution. Let C be the point on the circle between A and B such that $AM = AC$.

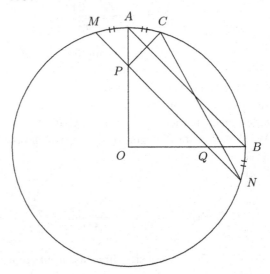

Since $\angle MAP = 45°$, we see that $MP \perp PC$. We also see that $AB = CN$. Using the Theorem of Pythagoras we see that

$$(AB)^2 = (CN)^2 = (CP)^2 + (PN)^2 = (MP)^2 + (PN)^2$$
$$= 56 + 144 = 200.$$

Therefore, $AB = 10\sqrt{2}$. Since triangle AOB is an isosceles right-angled triangle, we see that the radius $OA = AB/\sqrt{2} = 10$.

(283) *Problem.* A quadrilateral with sides of three, two, and four units in length (in that order) is inscribed in a circle with a diameter of five units.

What is the length of the fourth side of the quadrilateral?

Solution. The segments of length 2 and 4 can be swapped without changing the length of the fourth side.

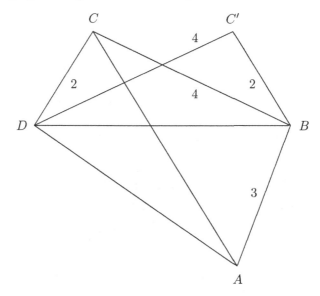

Let the quadrilateral be labelled $ABCD$ as shown above. If we now consider the diameter (length 5) which passes through A, we see that it must also pass through C forming a 3-4-5 triangle with the sides of length 3 and 4. But then triangle CDA must also be right-angled.

Thus, the fourth side has length $\sqrt{5^2 - 2^2} = \sqrt{21}$ units.

(284) *Problem.* Two circles, of radii R and r, $R > r$, are externally tangent to one another. Consider a common tangent of the two circles, not passing through their common point. Determine the maximal radius of a circle drawn in the region bounded by this tangent and the two circles.

Solution. Let O be the centre of the circle with radius R, and let P be the centre of the circle with radius r.

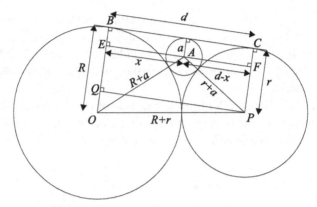

Let OB and PC be radii of these circles to the common tangent line, with B and C lying on the tangent line, as shown in the diagram above. Let A be the centre and a be the radius of the largest circle that can be drawn in the region bounded by the tangent line and the two given circles. Clearly, the largest such circle must be tangent (externally) to both of the given circles and also tangent to the common tangent of the two given circles, as shown in the diagram. Draw the radius AD of the new circle with D on the common tangent line. Since any tangent line to a circle is perpendicular to the radius of the circle at the point of contact, we see that OB, PC, and AD are all perpendicular to the tangent line, and thus are parallel to each other. Let d be the length of the line segment BC, and let x be the length of the line segment AE.

Let E and Q be points on OB such that $BDAE$ and $BCPQ$ are rectangles. Also let F be the point on PC such that $ADCF$ is a rectangle. We will now use the Theorem of Pythagoras on $\triangle POQ$, $\triangle AOE$, and $\triangle APF$ to obtain (respectively):

$$d^2 = (R+r)^2 - (R-r)^2 = 4rR,$$
$$x^2 = (R+a)^2 - (R-a)^2 = 4aR,$$
$$(d-x)^2 = (r+a)^2 - (r-a)^2 = 4ar.$$

From this we obtain

$$2\sqrt{rR} = d = x + (d-x) = 2\sqrt{aR} + 2\sqrt{ar}.$$

Solving for a we obtain

$$\sqrt{a} = \frac{\sqrt{rR}}{\sqrt{R} + \sqrt{r}}$$

or
$$a = \frac{rR}{R + r + 2\sqrt{rR}}.$$

(285) *Problem.* A semicircle is drawn with diameter AD. Point E is chosen on the semicircle, a perpendicular is dropped from E to AD intersecting AD at C. Assume that C is closer to the point D than to the point A. Now construct rectangle $ACEG$ and square $ABFG$, as shown in the diagram below. Let H be the intersection of CG and FB. Show that $FH = CD$.

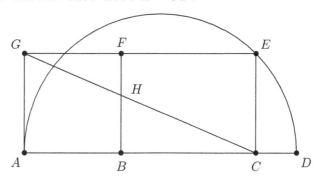

Solution. First draw the lines AE and ED. Since AD is the diameter of the semicircle, we see that $\angle ADE = 90°$.

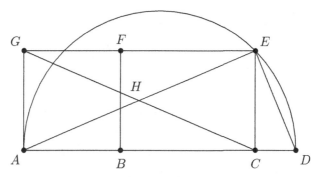

Since both triangles ACE and ECD are also right-angled we see that $\triangle ACE$ is similar to $\triangle ECD$. Furthermore, it can be seen that $\triangle ACE$ is similar to (indeed, it is congruent to) $\triangle GEC$, since $ACEG$ is a rectangle. It is also clear that $\triangle GEC$ is similar to $\triangle GFH$ since FB and EC are both perpendicular to AD. Thus, $\triangle ECD$ is similar to $\triangle GFH$. Since $ACEG$ is a rectangle, we have $EC = GA$, and since $ABFG$ is a square, we have $GF = GA$. Therefore, $GF = EC$, which means that $\triangle ECD$ is congruent to $\triangle GFH$. Thus, we have $FH = CD$, since they are corresponding parts in congruent triangles.

(286) *Problem.* The vertices of six squares coincide in such a way that they enclose triangles (see the picture below). Prove that the sum of the areas of the three outer squares (I, II, and III) equals three times the sum of the areas of the three inner squares (IV, V, and VI).

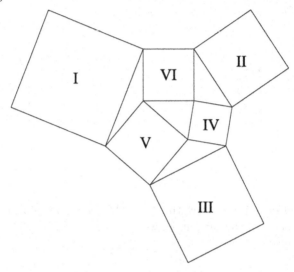

Solution. Let the figure be labelled as shown below.

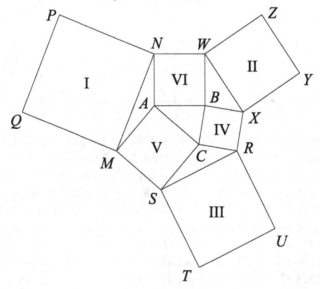

Let $x_1 = MN$, $x_2 = ZW$, $x_3 = RS$, $x_4 = BC$, $x_5 = CA$, $x_6 = AB$, $\alpha = \angle MAN$, $\beta = \angle WBX$, $\gamma = \angle RCS$, $A = \angle CAB$, $B = \angle ABC$, $C = \angle BCA$.

Then, we have $\alpha + A = \pi$, $\beta + B = \pi$, and $\gamma + C = \pi$. Using the Law of Cosines yields:

$$x_1^2 = x_5^2 + x_6^2 - 2x_5x_6 \cos \alpha,$$
$$x_2^2 = x_6^2 + x_4^2 - 2x_6x_4 \cos \beta,$$
$$x_3^2 = x_4^2 + x_5^2 - 2x_4x_5 \cos \gamma, \tag{1}$$
$$x_4^2 = x_6^2 + x_5^2 - 2x_6x_5 \cos A,$$
$$x_5^2 = x_4^2 + x_6^2 - 2x_4x_6 \cos B,$$
$$x_6^2 = x_5^2 + x_4^2 - 2x_5x_4 \cos C. \tag{2}$$

By adding the equations in (2) and simplifying, we have

$$x_4^2 + x_5^2 + x_6^2 = 2x_5x_6 \cos A + 2x_6x_4 \cos B + 2x_4x_5 \cos C$$
$$= -2x_5x_6 \cos \alpha - 2x_6x_4 \cos \beta - 2x_4x_5 \cos \gamma. \tag{3}$$

By adding the equations in (1), we have
$$x_1^2 + x_2^2 + x_3^2$$
$$= 2(x_4^2 + x_5^2 + x_6^2) - 2x_5x_6 \cos \alpha - 2x_6x_4 \cos \beta - 2x_4x_5 \cos \gamma,$$

and using (3), this becomes:
$$2(x_4^2 + x_5^2 + x_6^2) + x_4^2 + x_5^2 + x_6^2 = 3(x_4^2 + x_5^2 + x_6^2).$$

That is, Area of $(I + II + III) = 3 \times$ Area of $(IV + V + VI)$.

(287) *Problem.* Let $ABCD$ be a convex quadrilateral. Prove or disprove: there is a point E in the plane of $ABCD$ such that $\triangle ABE$ is similar to $\triangle CDE$.

(By a *convex* polygon we mean one in which the straight line segment joining any 2 interior points lies completely within the interior, that is, the polygon does not "fold in on itself".)

Solution. There always exists such a point E. If AB is parallel to CD, then choose E to be the point of intersection of the diagonals AC and BD (see the diagram on the next page).

Then $\angle ABE = \angle CDE$ and $\angle BAE = \angle DCE$, as they are alternate pairs of angles between the parallel lines AB and CD. This gives the required similarity.

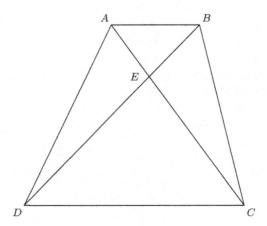

Next we suppose that AB and CD meet at F (see the diagram below). Construct the circumcircles of $\triangle ACF$ and of $\triangle DBF$. Let E be the second point of intersection of these circles. Because quadrilateral $AECF$ is cyclic we have $\angle EAF + \angle ECF = 180°$, which clearly implies that $\angle EAB = \angle ECD$. A similar argument shows that $\angle EDC = \angle EBA$. Thus, the required similarity is established for this case as well.

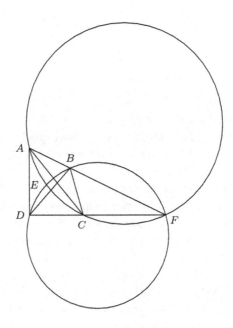

(288) *Problem.* Triangle PQR is isosceles, with $PQ = PR = 3$ and $QR = 2$, as shown below. The tangent to the circumcircle at Q meets (the extension of) PR at X, as shown. Find the length RX.

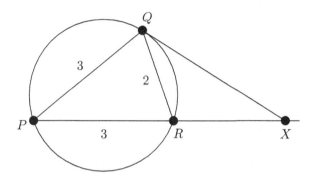

Solution. The chord QR subtends anywhere on the major arc of the circumference of the circle an angle equal in size to $\angle QPR$.

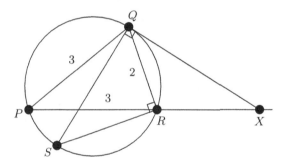

In particular, if QS is a diameter of the circle, then $\angle QSR = \angle QPR$. But $\angle QRS = 90°$, as is $\angle SQX = 90°$; thus $\angle QSR = \angle XQR$, whence $\angle QPR = \angle XQR$. With this fact in hand, and observing that $\angle X$ is a common angle, we note that triangles PQX and QRX are similar. Therefore,

$$\frac{PQ}{QX} = \frac{QR}{RX} \quad \text{and} \quad \frac{PX}{PQ} = \frac{QX}{QR}. \tag{1}$$

Using the first of these relations we have

$$\frac{3}{QX} = \frac{2}{RX},$$

$$QX = \frac{3}{2}RX.$$

Using the second relation in (1) together with this, we have

$$\frac{3 + RX}{3} = \frac{QX}{2} = \frac{\frac{3}{2}RX}{2},$$

$$6 + 2RX = \frac{9}{2}RX,$$

$$12 = 5RX.$$

Thus, $RX = 12/5 = 2.4$.

(289) *Problem.* $ABCD$ is a convex quadrilateral and E, F, G, H are the mid-points of AB, BC, CD, DA, respectively. Prove that the area of $ABCD$ is at most $EG \cdot FH$.

Solution.   Construct the two diagonals AC and BD. Let O be the point of intersection of AC and BD. Then join O to each of E, F, G, H.

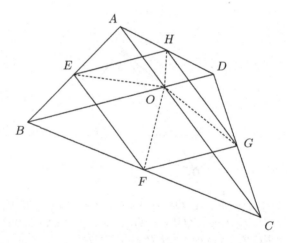

Next observe that $\triangle AEH$ and $\triangle OEH$ have the same area, since they have the same base, and (by virtue of the fact that E and H are mid-points of AB and AD, respectively) they have the same altitudes.

Similarly, $\triangle BEF$ and $\triangle OEF$ have the same area, as do $\triangle CFG$ and $\triangle OFG$, and $\triangle DGH$ and $\triangle OGH$.

Thus, the sum of the areas of triangles AEH, BEF, CFG, and DGH is equal to the area of the quadrilateral $EFGH$, which is thus exactly half of the area of quadrilateral $ABCD$.

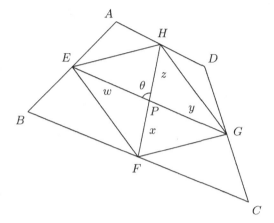

Let us now consider the diagonals EG and FH. Let P be the point of intersection of the diagonals, and let θ be the angle between these diagonals, as indicated in the above diagram. Further, let the lengths of PE, PF, PG, and PH be denoted by w, x, y, and z. Then the sum of the areas of the four triangles which make up quadrilateral $EFGH$ can be computed as

$$\frac{1}{2}xy\sin\theta + \frac{1}{2}yz\sin(\pi - \theta) + \frac{1}{2}zw\sin\theta + \frac{1}{2}wx\sin(\pi - \theta)$$

$$= \frac{1}{2}(xy + yz + zw + wx)\sin\theta$$

$$= \frac{1}{2}(x + z)(y + w)\sin\theta$$

$$= \frac{1}{2}FH \cdot EG\sin\theta \le \frac{1}{2}FH \cdot EG.$$

Since the area of $ABCD$ is twice that of $EFGH$, we see that the area of $ABCD$ is at most $FH \cdot EG$.

(290) *Problem.* P is a point inside a square $ABCD$ such that $PA = 1$, $PB = 2$, and $PC = 3$. What is the measure of $\angle APB$?

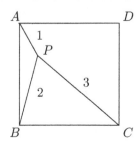

Solution. Rotate the entire figure by 90° centred at the point B. The square $ABCD$ becomes $A'BAD'$ under this rotation (see figure below).

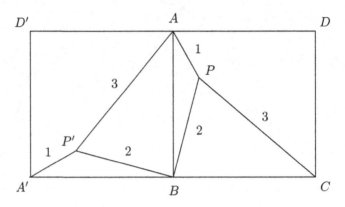

The segment BP is carried into the segment BP' to form the isosceles right-angled triangle BPP'. Thus, $\angle P'PB = 45°$. It remains for us to determine $\angle APP'$. By the Theorem of Pythagoras, we have $PP' = \sqrt{8}$, and then observing that $P'A = PC = 3$, we have

$$AP^2 + P'P^2 = 1 + 8 = 3^2 = P'A^2,$$

implying that $\angle APP'$ is a right-angle by the converse of the Theorem of Pythagoras (one could also use the Law of Cosines to justify this). Thus,

$$\angle APB = 90° + 45° = 135°.$$

Alternate Approach: Let a be the length of the side of the square $ABCD$. Applying the Law of Cosines to $\triangle ABP$, we get

$$\cos ABP = \frac{AB^2 + BP^2 - AP^2}{2 \cdot AB \cdot BP} = \frac{a^2 + 4 - 1}{4a} = \frac{a^2 + 3}{4a}.$$

If we apply the Law of Cosines to $\triangle BPC$, we obtain

$$\cos PBC = \frac{BP^2 + BC^2 - PC^2}{2 \cdot BP \cdot BC} = \frac{a^2 + 4 - 9}{4a} = \frac{a^2 - 5}{4a}.$$

Since $\angle ABP + \angle PBC = 90°$, we know that $\cos ABP = \sin PBC$.

Thus,

$$\cos^2 ABP = \sin^2 PBC = 1 - \cos^2 PBC,$$
$$\cos^2 ABP + \cos^2 PBC = 1,$$
$$\left(\frac{a^2 + 3}{4a}\right)^2 + \left(\frac{a^2 - 5}{4a}\right)^2 = 1,$$
$$a^4 + 6a^2 + 9 + a^4 - 10a^2 + 25 = 16a^2,$$
$$2a^4 - 20a^2 + 34 = 0,$$
$$a^4 - 10a^2 + 17 = 0,$$
$$a^2 = -\frac{10 \pm \sqrt{100 - 4 \cdot 17}}{2} = 5 \pm 2\sqrt{2}.$$

Since $\angle PBC \leq 90°$, we have $\cos PBC \geq 0$. If $a^2 = 5 - 2\sqrt{2}$, then (from above)

$$\cos PBC = \frac{a^2 - 5}{4a} = \frac{-2\sqrt{2}}{4(5 - 2\sqrt{2})} < 0,$$

which is impossible. Thus, we must have $a^2 = 5 + 2\sqrt{2}$. Applying the Law of Cosines yet again to $\triangle APB$, we have

$$\cos APB = \frac{AP^2 + BP^2 - AB^2}{2 \cdot AP \cdot BP} = \frac{1 + 4 - a^2}{4}$$
$$= \frac{5 - 5 - 2\sqrt{2}}{4} = -\frac{\sqrt{2}}{2},$$

from which we conclude that $\angle APB = 135°$.

(291) *Problem.* In quadrilateral $ABCD$ (shown below), we have $AB \parallel DC$, $BC = \sqrt{2}$ and $AB = AC = AD = \sqrt{3}$. Determine the length of BD.

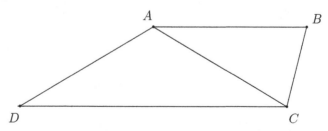

Solution. First draw the diagonal BD. Now let us denote the angles BAC, DAC, and ADB by x, y, z, respectively, as shown in the diagram on the next page.

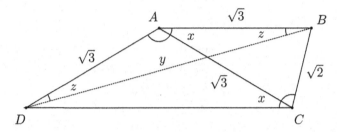

Since $AB \parallel DC$, we have $\angle ACD = x$ and $\angle ABD = z$. From (isosceles) $\triangle ACD$, we conclude that $2x + y = 180°$, or (equivalently) $x + y = 180° - x$. From (isosceles) $\triangle ABD$, we similarly conclude that $2z + x + y = 180°$, which means that $z = x/2$.

Next we note (by dropping a perpendicular from A to BD) that:

$$\sin\left(\frac{x+y}{2}\right) = \frac{\frac{1}{2}BD}{\sqrt{3}},$$

which means that

$$BD = 2\sqrt{3}\sin\left(\frac{x+y}{2}\right). \tag{1}$$

Applying the Law of Sines to (isosceles) $\triangle ABC$ we have:

$$\frac{BC}{\sin x} = \frac{AB}{\sin\frac{1}{2}(180° - x)},$$

$$\frac{\sqrt{2}}{\sin x} = \frac{\sqrt{3}}{\sin\frac{1}{2}(x+y)},$$

$$\sin\left(\frac{x+y}{2}\right) = \frac{\sqrt{3}}{\sqrt{2}}\sin x.$$

Using this in (1) we have

$$BD = 2\sqrt{3}\frac{\sqrt{3}}{\sqrt{2}}\sin x = 3\sqrt{2}\sin x. \tag{2}$$

Since $z = x/2$ we also have $\sin x = 2\sin z \cos z$. By dropping a perpendicular from A to BC we compute

$$\sin z = \frac{\frac{1}{2}\sqrt{2}}{\sqrt{3}},$$

$$\cos z = \frac{\sqrt{\frac{5}{2}}}{\sqrt{3}}.$$

Putting this into (2), we obtain

$$BD = 3\sqrt{2}\cdot 2\frac{\frac{1}{2}\sqrt{2}}{\sqrt{3}}\frac{\sqrt{\frac{5}{2}}}{\sqrt{3}} = \frac{3\sqrt{2}\sqrt{2}\sqrt{5}}{\sqrt{3}\sqrt{3}\sqrt{2}} = \sqrt{10}.$$

(292) *Problem.* The diagonals of a square $ABCD$ meet at E, and the bisector of $\angle CAD$ crosses DE at G and DC at F. If the length of GE is 24, how long is FC?

Solution. In $\triangle ADF$, we have $\angle DAF = 22\frac{1}{2}^{\circ}$, whence, $\angle DFA = 67\frac{1}{2}^{\circ}$.

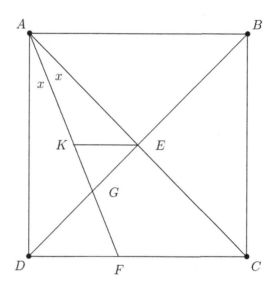

Now construct the line through E parallel to CD and meeting AF at K. Clearly, $\angle GEK = \angle GDF = 45^{\circ}$ and $\angle GKE = \angle GFD = 67\frac{1}{2}^{\circ}$. Thus, the third angle of $\triangle GEK$ is $\angle EGK = 180^{\circ} - 45^{\circ} - 67\frac{1}{2}^{\circ} = 67\frac{1}{2}^{\circ}$. This means that $\triangle EGK$ is isosceles with $KE = GE = 24$. Now triangles AKE and AFC are obviously similar, which implies that

$$\frac{FC}{KE} = \frac{AC}{AE} = \frac{2}{1}.$$

Therefore, $FC = 2 \cdot KE = 48$.

(293) *Problem.* Let us consider a convex pentagon $ABCDE$ with $AB = AE = CD = 1$, $\angle ABC = \angle DEA = 90^{\circ}$, and $BC + DE = 1$. Find the area of the pentagon.

Solution. First drop a perpendicular from A to CD meeting CD at the point F. Let $\alpha = DE$; then $BC = 1 - \alpha$. Similarly, let $\beta = DF$; then $CF = 1 - \beta$.

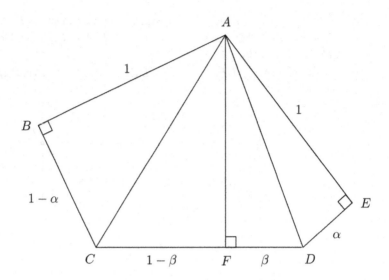

Applying the Theorem of Pythagoras four times yields

$$1 + \alpha^2 = AD^2 = AF^2 + \beta^2$$

and

$$1 + (1 - \alpha)^2 = AC^2 = AF^2 + (1 - \beta)^2.$$

Rearranging and combining these gives us

$$1 = AF^2 + \beta^2 - \alpha^2 = 1 + 2\beta - 2\alpha.$$

We conclude that $\beta = \alpha$, which further implies that $AF = 1$. Thus, the four right-angled triangles which make up the pentagon $ABCDE$ all have altitude 1. This means that the area is one half of the sum of the bases; that is, $\frac{1}{2}(\alpha + \beta + 1 - \beta + 1 - \alpha) = 1$.

(294) *Problem.*   Given a rectangle $ABCD$ with $AB = CD > AD = BC$, construct point X, Y on CD between C and D such that $AX = XY = YB$.

Solution.   **Construction.**   Let M be the mid-point of the side CD.

Take a point P on the ray MA such that $DP = DC$. Take a point X on the segment DM such that $AX \parallel PD$. Take a point Y on the segment MC such that $MX = MY$. Then X and Y are the required points.

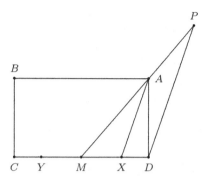

Proof. Since $AX \parallel PD$ and $DP = DC$,

$$AX : XM = PD : DM = CD : DM = 2 : 1.$$

Therefore, $AX = 2XM = XY$, and by symmetry $AX = BY$.

Alternatively:

Let M be the mid-point of CD. Extend segment AM beyond M to the point O such that

$$AO = \left(\frac{4}{3}\right) AM.$$

The circle Ω of centre O and radius $(2/3)AM$ is the locus of points whose distance from A equals twice the distance from M (the Apollonius circle). Since $AD < CD = 2MD$, point D lies outside Ω. Hence, Ω intersects segment MD; the point of intersection is the desired X. Y shall lie symmetrically to X with respect to M.

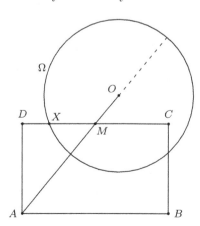

(295) *Problem.*   In parallelogram $ABCD$, angle A is acute and $AB = 5$. Point E is on AD with $AE = 4$ and $BE = 3$. The line through B perpendicular to CD intersects CD at F. If $BF = 5$, find EF. We are looking for a geometric solution here, not a solution involving trigonometry!

Solution.   Let P be the point on the segment AE with $AP = 3$.

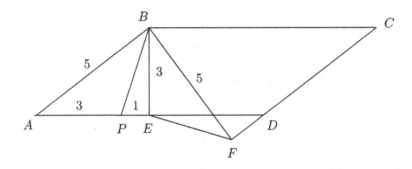

Since ABE is a 3–4–5 triangle, the converse of the Theorem of Pythagoras implies that $BE \perp AD$. We also have $AB \perp BF$.

Since $\angle BAE + \angle ABE = 90°$ and $\angle ABE + \angle EBF = 90°$, we see that $\angle BAP = \angle BAE = \angle EBF$.

Therefore, triangles BAP and FBE are congruent (SAS). Therefore, $EF = PB$ which (by the Theorem of Pythagoras) is $\sqrt{1^2 + 3^2} = \sqrt{10}$.

(296) *Problem.*   In the figure below, the measures of the four angles labelled α, β, γ, and δ are given. Determine the measure of angle EDC in terms of α, β, γ, and δ.

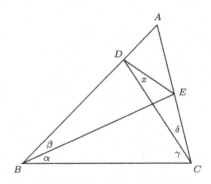

Solution. Applying the Law of Sines to $\triangle CDE$, we get

$$\frac{CD}{CE} = \frac{\sin\left(\pi - (x + \delta)\right)}{\sin x} = \frac{\sin(x + \delta)}{\sin x}. \tag{1}$$

Doing the same for $\triangle BCD$ and $\triangle BCE$, we obtain

$$\frac{CD}{CB} = \frac{\sin(\alpha + \beta)}{\sin\left(\pi - (\alpha + \beta + \gamma)\right)} = \frac{\sin(\alpha + \beta)}{\sin(\alpha + \beta + \gamma)}, \tag{2}$$

$$\frac{CB}{CE} = \frac{\sin\left(\pi - (\alpha + \gamma + \delta)\right)}{\sin \alpha} = \frac{\sin(\alpha + \gamma + \delta)}{\sin \alpha}. \tag{3}$$

Using (1), (2), and (3), we see that

$$\frac{\sin(x + \delta)}{\sin x} = \frac{\sin(\alpha + \beta)}{\sin(\alpha + \beta + \gamma)} \cdot \frac{\sin(\alpha + \gamma + \delta)}{\sin \alpha}. \tag{4}$$

Since $\sin(x + \delta) = \sin x \cos \delta + \cos x \sin \delta$, we see that (4) can be reduced to

$$\cos \delta + \cot x \sin \delta = \frac{\sin(\alpha + \beta) \sin(\alpha + \gamma + \delta)}{\sin \alpha \sin(\alpha + \beta + \gamma)},$$

or

$$\cot x = \frac{\sin(\alpha + \beta) \sin(\alpha + \gamma + \delta)}{\sin \alpha \sin \delta \sin(\alpha + \beta + \gamma)} - \cos \delta,$$

from which x can be determined uniquely, since $0 < x < \pi$.

(297) *Problem.* Given any four distinct points on a line, construct a square by drawing two pairs of parallel lines through the four points.

Solution. Let A, B, C, D be the four points on the line (in that order).

Solution I: Select the two segments AB and CD, and erect perpendiculars BA' and CD' on the same side of the line such that $BA' = CD$ and $CD' = AB$ (as in the figure below). We then draw the lines AA' and DD' and complete the construction by drawing the necessary parallel lines.

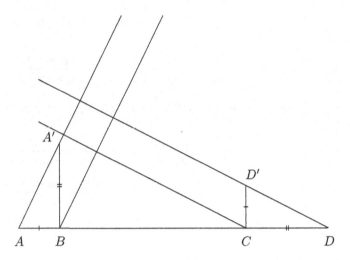

By using similar triangles, we can see that the angles of the resulting parallelogram are all right angles (making it a rectangle), and that the length of each side of this rectangle is given by

$$s = \sqrt{AB \cdot BA'} = \sqrt{AB \cdot CD},$$

which means that it is square.

Solution II: Construct the circles C_1 and C_2 having AB and CD as diameters, respectively. Construct the points E and F on C_1 and C_2, respectively, such that E and F are on opposite sides of the given line containing our four points and such that $AE = EB$ and $CF = FD$.

The line EF meets the circle C_1 at a second point, say G, and also meets the circle C_2 at a second point, say H. Let O be the point of intersection of EF and the given line of four points.

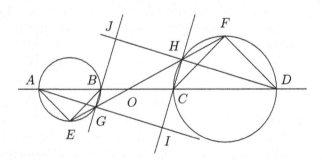

Note that $\angle AGE = \angle ABE = 45°$ and $\angle DHF = \angle DCF = 45°$. Now, draw the lines AG, BG, CH, and DH. Let I be the intersection of AG and CH, and let J be the intersection of BG and DH.

We will now show that $GIHJ$ is a square. Since $\angle AGB$ and $\angle CHD$ are both angles subtended by a diameter, they are both right angles.

Since $\angle AGE = 45°$, so is $\angle HGI = 45°$ (because $\angle JGI = 90°$). We also have $\angle GHI = 45°$, which means that $\angle GIH = 90°$ (using $\triangle GHI$).

Thus, GI is perpendicular to HI. Similarly, GJ is perpendicular to HJ. This means that $GIHJ$ is a rectangle.

However, $\triangle GHI$ is an isosceles triangle, since $\angle HGI = \angle GHI = 45°$, implying that $GI = HI$, which shows that $GIHJ$ is a square.

(298) *Problem.* The quadrilateral $ABCD$ in the figure below has base BC of 6 cm, a right angle at B, and $\angle BCD = 135°$. The altitude AE drawn from the vertex A to the side CD has length 12 cm, and the length of ED is 5 cm. Find the area of the quadrilateral $ABCD$.

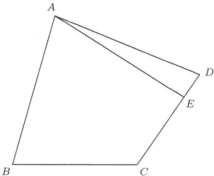

Solution. We see that the area of triangle AED is 30 cm^2.

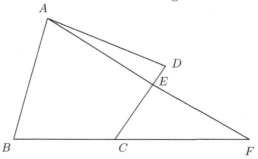

Next we will extend line segments AE and BC to intersect in the point F. Clearly, both triangles ABF and CEF are isosceles right triangles. If we let the side CE have length x cm, then EF also has length x cm, and by applying the Theorem of Pythagoras to triangle CEF we see that CF has length $x\sqrt{2}$ cm. Because triangle ABF is an isosceles right triangle, we have

$$12 + x = \sqrt{2}(6 + x\sqrt{2}),$$
$$\text{or} \quad x = 12 - 6\sqrt{2}.$$

Now the area A of the quadrilateral $ABCD$ can be decomposed into triangles as follows:

$$
\begin{aligned}
A &= \triangle ABF - \triangle CEF + \triangle AED \\
&= \frac{1}{2}\left[6 + \sqrt{2}(12 - 6\sqrt{2})\right]^2 - \frac{1}{2}\left[12 - 6\sqrt{2}\right]^2 + 30 \\
&= 18\left[1 + \sqrt{2}(2 - \sqrt{2})\right]^2 - 18\left[2 - \sqrt{2}\right]^2 + 30 \\
&= 18\left[2\sqrt{2} - 1\right]^2 - 18\left[2 - \sqrt{2}\right]^2 + 30 \\
&= 18(8 - 4\sqrt{2} + 1) - 18(4 - 4\sqrt{2} + 2) + 30 \\
&= 18(9 - 6) + 30 = 84.
\end{aligned}
$$

(299) *Problem.* Circles of radius 3 and 6 are externally tangent to each other and are internally tangent to a circle of radius 9. The circle of radius 9 has a chord that is the common external tangent of the other two circles.

Find the square of the length of this chord. (See the diagram on the next page.)

Solution. Let the centres of the circles of radius 3, 6, and 9 be denoted by A B, and C, respectively.

Draw the lines through each of these centres which are perpendicular to the chord, meeting the chord at D, E, and F, respectively. Clearly AD has length 3 as it is the radius of the small circle; similarly BE has length 6.

Let P be the point of intersection of the chord with the common diameter of the three circles.

Since the lines AD and BE are parallel, since BE is twice as long as AD and since the distance between them (measured along the common diameter of all three circles) is 9 units, we see that the distance from A to P is also 9 units.

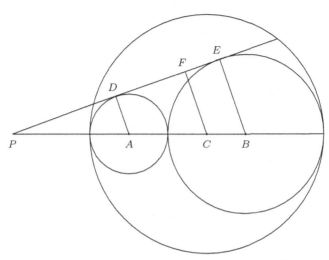

By using similar triangles we can conclude that the distance CF is 5 units. Now by using the Theorem of Pythagoras on the right-angled triangle formed by C, F and one end of the chord, we see that the square of half the length of the chord is $81 - 25 = 56$. Thus, the square of twice this length (that is, the square of the length of the chord itself) is $4 \times 56 = 224$.

(300) *Problem.*   Three circles have the same radius r and pass through the same point O. Let A, B, and C be the other points of inter-section of the three pairs of these circles, respectively.

Prove that A, B, and C lie on a circle of radius r.

Solution.   Let D be the centre of the circle containing the points O, A, and B; let E be the centre of the circle containing O, A, and C; and let F be the centre of the circle containing O, B, and C. First consider the quadrilateral $OEAD$: since all 4 sides are radii of the circle centred either at D or at E, we see that it is a rhombus, which implies that OD is equal in length and parallel to EA and that OE is equal in length and parallel to DA. Similarly we can use the quadrilateral $OECF$ to establish that OE is equal in length and parallel to FC and OF is equal in length and parallel to EC, and quadrilateral $ODBF$ to show that OD is equal in length and parallel to FB and OF is equal in length and parallel to DB.

All of the above line segments have length r. From the above, we can deduce that EA is equal in length and parallel to FB, DA is equal in length and parallel to FC, and DB is equal in length and parallel to EC. Now define the point G to be the point which "completes the parallelogram" having adjacent sides DA and DB. Since both DA and DB have length r, so do AC and BG. Since AC is equal in length and parallel to DB, it is also equal in length and parallel to EC, which means that the quadrilateral $GAEC$ is a rhombus, and thus, CG is equal in length and parallel to EA. Thus, G is at a distance r from each of A, B, and C. Thus, a circle centred at G of radius r will pass through A, B, and C.

(301) *Problem.* A hexagon inscribed in a circle has three consecutive sides of length a and three consecutive sides of length b. Determine the radius of the circle in terms of a and b.

Solution. Let R be the radius of the circle. Note that each side of length a subtends the same angle at the centre of the circle, and similarly for the sides of length b. Thus, we can rearrange the sides so that they alternate from a to b and still have a hexagon inscribed inside the same circle. Now each adjacent pair of interior angles of the hexagon is now made up by putting together two isosceles triangles, one with base a and one with base b and with equal sides of length R for both, we see that the sum of each pair of interior angles has the same measure, namely $120°$.

Let A, B, C, D, E, F be the labels in order of the vertices of the hexagon, and let O be the centre of the circle. Consider the chord AC. Applying the Law of Cosines to the triangles ABC and AOC, and recalling that $\cos 120° = -1/2$, we get

$$(\overline{AC})^2 = a^2 + b^2 - 2ab\cos 120° = a^2 + b^2 + ab,$$

and $(\overline{AC})^2 = R^2 + R^2 - 2R \times R\cos 120° = 2R^2(1 - \cos 120°) = 3R^2$,

from which it follows that

$$3R^2 = a^2 + b^2 + ab, \quad \text{so that,} \quad R = \sqrt{\frac{a^2 + b^2 + ab}{3}}.$$

(302) *Problem.* In a parallelogram $ABCD$, the bisector of angle ABC intersects AD at P. If $PD = 5$, $BP = 6$ and $CP = 6$, find AB.

Solution. Let $\angle PBC = x$. Then, since BP is the angle bisector of angle ABC, we see that $\angle ABP = \angle PBC = x$. Since $\triangle BPC$ is isosceles we see that $\angle PBC = \angle PCB = x$.

Also, $\angle PCB = \angle CPD = x$, since they are alternate interior angles for the parallel lines BC and PD.

This forces $\angle APB = x$, since $\angle BPC = 180° - 2x$, and A, P, D are collinear.

Thus, $\triangle ABP$ is isosceles and it is similar to $\triangle PBC$. Now let $AB = y$. By similar triangles, we have

$$\frac{AB}{BP} = \frac{BP}{BC} = \frac{BP}{AP + PD},$$
$$\frac{y}{6} = \frac{6}{y + 5},$$
$$y^2 + 5y - 36 = 0,$$
$$y = \frac{-5 \pm \sqrt{25 + 144}}{2} = \frac{-5 \pm 13}{2} = 4 \text{ or } -9.$$

Since the distance cannot be negative, we get $AB = 4$.

(303) *Problem.* In the acute-angled triangle, ABC, let H be its orthocentre (that is, the point of intersection of the three altitudes), and let O be its circumcentre (that is, the centre of the circle passing through all three vertices of the triangle; such a circle is also called the circumcircle). Line BO is extended to meet the circumcircle at D.

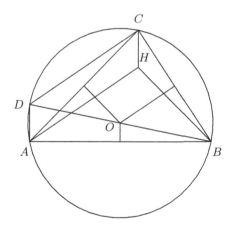

Show that $ADCH$ is a parallelogram.

Solution. Extend AH and CH to meet BC and AB at E and F, respectively.

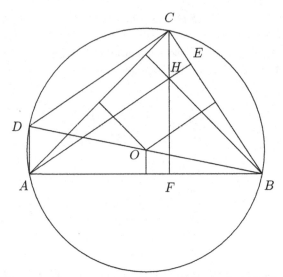

Since BD is a diameter of cyclic quadrilateral $ABCD$, we have $BC \perp CD$, and since $BC \perp AE$, we have $AE \parallel CD$. Similarly, $DA \parallel CF$, so that $ADCH$ is indeed a parallelogram.

(304) *Problem.*   In the right triangle ABC, let $AB = c$, $BC = a$, and $AC = h$. If $AE = EF = FC = h/3$, and if $\angle EBF = \alpha$, show that

$$\tan \alpha = \frac{3ac}{2h^2}.$$

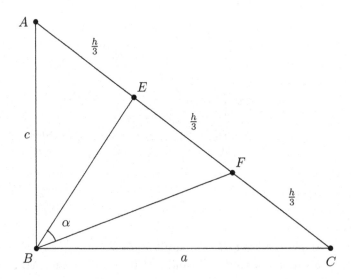

Solution.   Jim Totten thought that this was the nicest solution of those he had seen.

Extend the triangle to the parallelogram $ABCD$ and extend the lines BE and BF to meet AD and CD at G and H, respectively.

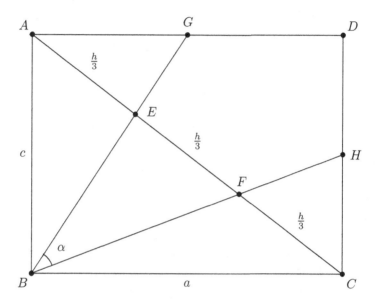

By considering the similar triangles BEC and AEG, we can deduce that G is the mid-point of the line AD.

Similarly, H is the mid-point of CD.

Now by looking at the right triangles ABG and BCH, we obtain:

$$\tan ABG = \frac{a/2}{c} = \frac{a}{2c}$$

$$\text{and}\quad \tan HBC = \frac{c/2}{a} = \frac{c}{2a}.$$

We now must consider the trigonometric identities:

$$\tan \alpha \tan(ABG + HBC) = 1$$

$$\text{or}\quad \tan(ABG + HBC) = \frac{\tan ABG + \tan HBC}{1 - \tan ABG \tan HBC},$$

from which the result follows.

(305) *Problem.*   A trapezoid is divided into four triangles by its diagonals. Let A and B denote the areas of the triangles adjacent to the parallel sides. Find, in terms of A and B, the area of the trapezoid.
Solution.   Denote the areas of the four triangles (into which the trapezoid is divided by the diagonals) by A, B, C, D as in the figure.

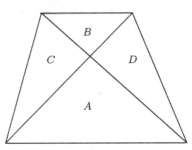

If one diagonal divides the other in the ratio $u : v$, then
$$\frac{C}{B} = \frac{A}{D} = \frac{u}{v},$$
so that $A \times B = C \times D$. Since triangles with the same base and between the same parallels have equal area, we have $A+C = A+D$, whence $C = D = \sqrt{AB}$. Thus, the area of the trapezoid is
$$A + B + C + D = A + B + 2\sqrt{AB} = (\sqrt{A} + \sqrt{B})^2.$$

(306) *Problem.*   A large cylindrical pot has been set over an open fire on crosspieces of metal bars, at right angles to each other and with sharp upper edges marked in inches measured out from their point of intersection, for some forgotten reason.
The pot has been pushed to a precarious position, and from where we are we can see that the circular bottom just intersects the bars at the 6-, 8-, and 15-inch marks at the left, front, and right, respectively. What is the diameter of the pot?
Solution.   Set up a coordinate system with the intersection of the cross pieces as the origin. Let (a, b) be the coordinates of the centre of the circular base of the pot. The points of intersection of the circular base and the cross pieces are $(-6, 0)$, $(0, -8)$ and $(15, 0)$. Since the distance from the centre to each of these points is the same, we can use the distance formula twice to obtain 2 equations in a and b. These equations can be solved to get $a = 4.5$ inches and $b = 1.625$ inches, from which we can calculate the diameter of the pot to be 21.25 inches.

(307) *Problem.*   Inscribe a rectangle of base b and height h and an isosceles triangle of base b in a circle of radius one as shown.

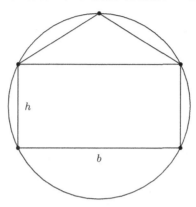

For what values of h do the rectangle and triangle have the same area?

Solution.   Draw a diameter vertically through the apex of the triangle. Since a second congruent triangle could have been drawn below the rectangle, the altitude of the triangle is $\frac{1}{2}(2 - h)$. Since we have equal areas, we obtain

$$\left(\frac{1}{2}b\right)\left(\frac{1}{2}\right)(2 - h) = bh$$

$$\text{or} \qquad h = \frac{2}{5}.$$

(308) *Problem.*   In the diagram below, show that the two differently shaded regions in the figure have the same area.

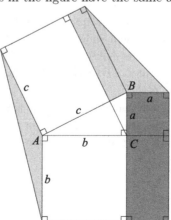

Solution. First of all we will label the diagram as shown below:

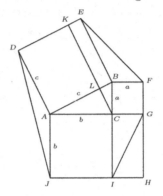

We will also draw the line GI which divides the rectangle $CGHI$ into the two congruent triangles CGI and HIG.

Now consider the triangles ABC and ADJ. Pivot the triangle ADJ clockwise by $90°$. Then the point D will coincide with the point B, and the line segment AJ will simply be an extension of the line segment CA.

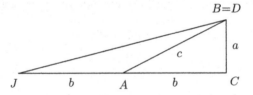

Since AJ and AC have the same length, and since the heights of both triangles are the same, their areas are the same. (Note that this argument does NOT require $\triangle ABC$ to be right-angled.)

A similar argument will show that $\triangle BEF$ and $\triangle CGI$ must also have the same area as $\triangle ABC$. Thus, we have established that the sum of the areas of $\triangle ADJ$ and $\triangle BEF$ is the same as the area of the rectangle $CGHI$.

It remains to show that the area of the square $BCGF$, namely a^2, is the same as the area of the rectangle $BEKL$. Since $\triangle CLB$ and $\triangle ABC$ are both right-angled and have common angle B, they are similar. Setting $x = BL$ we then get

$$\frac{x}{a} = \frac{a}{c},$$

which can be rewritten as $cx = a^2$. However, the area of the rectangle $BEKL$ is cx, which means that it has the same area as the square $BCGF$, and we are done.

(309) *Problem.* Let C be a circle in the plane. Let C_1 and C_2 be non-intersecting circles touching C internally at points A and B, respectively.

Let t be a common tangent of C_1 and C_2, touching them at points D and E, respectively, such that both C_1 and C_2 are on the same side of t. Let F be the point of intersection of AD and BE.

Show that F lies on C.

Solution. Let O, O_1, O_2 be the centres of the circles C, C_1, C_2, respectively.

Let G be the point of intersection between the circle C and the line AD, and let H be the point of intersection between the circle C and the line BE.

We must show that $F = G = H$, which follows if we simply show that $G = H$.

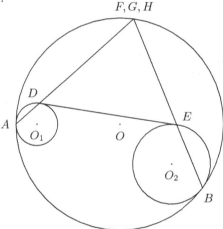

Clearly triangles ADO_1 and AGO are isosceles and (since O_1 lies on OA) we see that 0_1D and OG are parallel. Similarly we see that O_2E and OH are parallel. Since the radius of a circle is perpendicular to the tangent at the point of tangency, we see that O_1D and O_2E are both perpendicular to t and hence, parallel to each other. This means that OG and OH are parallel to each other. Since they have a point in common, they must be the same line, which means that $G = H$, since both of them lie on both the line and the circle C.

(310) *Problem.* Given five distinct points P_1, P_2, P_3, P_4, P_5 in the plane having integer coordinates, prove that there is at least one pair (P_i, P_j) with $i \neq j$ such that the line $P_i P_j$ contains a point Q having integer coordinates and lying strictly between P_i and P_j.

Solution. The points can be characterized according to the parity of their x and y coordinates. There are only four such classes: (even, even), (even, odd), (odd, even), and (odd, odd).

Since we are given five such points, at least two must have the same parity of coordinates by the Pigeon-hole Principle. Suppose they are P_i and P_j, where $P_i = (x_i, y_i)$ and $P_j = (x_j, y_j)$. Then $x_i + x_j$ is even and $y_i + y_j$ is even, since the x_i, x_j have the same parity and the y_i, y_j have the same parity. Hence the mid-point

$$Q = \left(\frac{x_i + x_j}{2}, \frac{y_i + y_j}{2} \right)$$

has integer coordinates and lies strictly between P_i and P_j.

(311) *Problem.* The parabola $y = 4x^2 - 24x + 31$ crosses the x-axis at $E_1 = (e_1, 0)$ and $E_2 = (e_2, 0)$, both to the right of the origin. A circle also passes through these two points. Find the length of the tangent from the origin to any such circle.

Solution. We will do the problem in general. Let the parabola be $y = ax^2 - 2bx + c$. By completing the square we have

$$y = a \left(x^2 - \frac{2b}{a}x + \left(\frac{b}{a} \right)^2 \right) + c - \frac{b^2}{a} = a \left(x - \frac{b}{a} \right)^2 + \frac{ac - b^2}{a}.$$

Now, the parabola is symmetric about the line $x = b/a$. If a and c are positive and if $ac < b^2$, then the parabola crosses the x-axis at two points; and if b is positive, both points will lie to the right of the origin; in fact from the above or by using the quadratic formula we see that these points have x-coordinates:

$$e_1 = \frac{b + \sqrt{b^2 - ac}}{a} \quad \text{and} \quad e_2 = \frac{b - \sqrt{b^2 - ac}}{a}.$$

Let the circle in the problem have centre $C = (h, k)$ and radius r. Since the centre of the circle must lie on the axis of symmetry for the parabola, we see that $h = b/a$.

Drop a perpendicular from C to the x-axis meeting it at D. Then the distance from D to E_1 is

$$b - c_1 = \frac{\sqrt{b^2 - ca}}{a}.$$

Applying the Theorem of Pythagoras to triangle CDE, we see that

$$r^2 = k^2 + \frac{b^2 - ac}{a^2}. \tag{1}$$

Let T be the point of tangency for the line from the origin to the circle, and let d be the length of the tangent. Applying the Theorem of Pythagoras to triangle OCD and OCT, we have

$$(OC)^2 = \left(\frac{b}{a}\right)^2 + k^2 \quad \text{and} \quad (OC)^2 = r^2 + d^2.$$

Equating these, solving for d^2, and using (1), yields

$$d^2 = \left(\frac{b}{a}\right)^2 + k^2 = \frac{b^2}{a^2} - \frac{b^2 - ac}{a^2} = \frac{c}{a},$$

whence, the length of the tangent line d is $\sqrt{c/a}$.

In our particular problem we have $a = 4$, $b - 12$, and $c = 31$, so that the length of the tangent line is $\sqrt{31/4} = \sqrt{31}/2$.

Chapter 8

Probability

(312) *Problem.* Max and his wife Min each toss a pair of dice to determine where they will spend their vacation. If either of Min's dice displays the same number of spots as either of Max's, she wins and they go to Paris. Otherwise, they go to Banff. What is the chance that they will see Sulphur Mountain this year?

Solution. If Max throws doubles, the probability of Min not matching his number with either of his dice is $(\frac{5}{6})^2$. If his two dice come up with different numbers, then the probability is $(\frac{4}{6})^2$. Now given the value of Max's first die, his chance of having the second matching it is $\frac{1}{6}$, and of having it different is thus $\frac{5}{6}$. Thus, the probability of going to Banff is

$$\frac{1}{6} \cdot \left(\frac{5}{6}\right)^2 + \frac{5}{6} \cdot \left(\frac{4}{6}\right)^2 \approx 0.486.$$

Since Sulphur Mountain is in Banff, the probability they will see it is 0.486.

(313) *Problem.* If the odds against a horse winning a race are $a : b$, then we may call $\frac{b}{a+b}$ the apparent chance of that horse winning the race. Prove that, if the sum of the apparent chances of winning of all the horses in a race is less than 1, one can arrange bets so as to make sure of winning the same sum of money whatever the outcome of the race.

Solution. Let there be n horses in the race. If the bettor bets $\frac{b_k}{a_k+b_k}$ against $\frac{a_k}{a_k+b_k}$ that the kth horse will win (for $1 \leq k \leq n$), then, if the kth horse does win, the bettor wins $\frac{a_k}{a_k+b_k}$ and loses

$$\sum_{\substack{i=1 \\ i \neq k}}^{n} \frac{b_i}{a_i + b_i} = \sum_{i=1}^{n} \frac{b_i}{a + i + b_i} - \frac{b_k}{a_k + b_k}.$$

Therefore, the bettor's winnings are:

$$\frac{a_k}{a_k + b_k} - \sum_{i=1}^{n} \frac{b_i}{a_i + b_i} + \frac{b_k}{a_k + b_k} = 1 - \sum_{i=1}^{n} \frac{b_i}{a_i + b_i},$$

which is positive and the same whatever horse wins.

(314) *Problem.*   In a certain class 25% of the students are blue-eyed and 75% are brown-eyed. If 10% of the blue-eyed students are left-handed, and 5% of the brown-eyed students are left-handed, what percentage of the left-handed students are blue-eyed?

Solution.   Let the number of students be n. Then there are $0.25n$ blue-eyed students and $0.75n$ brown-eyed students. The number of blue-eyed students who are left-handed is $0.10(0.25n) = 0.025n$. The number of brown-eyed students who are left-handed is then $0.05(0.75n) = 0.0375n$. Thus, the total number of left-handed students is $0.025n + 0.0375n = 0.0625n$. In order to compute the percentage of left-handed students who are blue-eyed we must evaluate:

$$\frac{\#\text{Blue-eyed, left-handed} \times 100}{\#\text{Left-handed}} = \frac{0.025n}{0.0625n} \times 100 = 40\%.$$

(315) *Problem.*   If two committees of three persons each are to be selected at random from an executive of nine persons, what is the probability that the oldest person on one committee is younger than the youngest person on the other committee.

Solution.   We need only consider the 6 people selected to be on committees, and we can represent these people (A through F) by their ages. Assume that $A > B > C > D > E > F$. The problem then becomes what is the probability that D, E, and F are on the same committee, since then the eldest (D) would be younger than the youngest on the other committee, namely, C.

There are $\binom{6}{3} = 20$ possible committees and only one of these would consist of D, E, and F. Hence, the desired probability is $\frac{1}{20} = 0.05$.

(316) *Problem.*   Abner Appleby is scheduled to play a two game chess match with Barney Blitz. If they are tied after two games, there will be a play-off, and the first player to win a game thereafter wins the match. The situation does not look too promising for Abner, since Barney is the stronger player.

Abner can play a daring game, which he has a 45% probability of winning and a 55% probability of losing; or he can play a conservative game, which he has a 90% probability of drawing and a 10% probability of losing.

Either way he would wind up ten games behind out of every hundred played (this is based on awarding 1 point for a win, half a point for a draw and 0 points for a loss, as is the standard scoring system for chess matches).

What is Abner's best strategy, and what is the probability of his winning the match?

Solution. If he has any hope of winning, he must play a daring strategy whenever he is tied or is losing.

It can also be shown that if he is ahead, he should play a conservative strategy.

(This seems intuitively clear, but can be shown as a sound strategy mathematically.)

Thus, there are four possible outcomes after two games:

1. L-L: Abner loses match
- probability is $(0.55) \times (0.55) = 0.3025$.
2. L-W: tie-breaker required
- probability is $(0.55) \times (0.45) = 0.2475$.
3. W-L: tie-breaker required
- probability is $(0.45) \times (0.10) = 0.0450$.
4. W-D: Abner wins match
- probability is $(0.45) \times (0.90) = 0.4050$.

The overall probability of a tie-breaker being required is 0.2925, and with Abner's strategy he must play a daring game.

Thus, in the event of a tie-breaker he has a probability of 0.45 of winning.

His overall chances of winning are:

$$0.4050 + (0.2925 \times 0.45) = 0.4050 + 0.131625 = 0.536625,$$

which is almost a 54% chance of winning.

(317) *Problem.* The probability that an old-fashioned watch will stop is 10% per hour. If the watch is set and started at 12 o'clock, what is the probability (to the nearest 1%) that the hour hand will point below the horizontal 12 hours later?

Solution. If the hour hand stops below the horizontal in the first 12 hours, it must have stopped during hour n where n is at least 4 and at most 9. That is, the watch is still running after 3 hours, but has stopped running before 9 hours. The probability that we want is then the probability that the watch is still running after 3 hours less the probability that it is still running after 9 hours, that is $p = (0.9)^3 - (0.9)^9 = 0.3416$. Thus, to the nearest percent, the probability is 34%.

(318) *Problem.*   The Miracle Marble Manufacturing Company manufactures orange marbles and purple marbles. A bag of their marbles may contain any combination of orange and purple marbles (including all orange or all purple) and all combinations are equally probable.

Henry bought a bag of their marbles and pulled one out at random. It was purple. What is the probability that if he pulled out a second marble at random it would also be purple?

Solution.   There is a probability of $\frac{2}{3}$ that the second marble would also be purple. This result is independent of the number of marbles in the bag. The proof is as follows:

Let n be the number of marbles in the bag.

Each of the following $n + 1$ combinations has a probability of $\frac{1}{n+1}$.

Orange	Purple
0	n
1	$n - 1$
2	$n - 2$
$\vdots$	$\vdots$
$n - k$	k
$\vdots$	$\vdots$
n	0

There are a total of $\frac{n(n+1)}{2}$ purple marbles in all of the above combinations, each of which has an equal probability, $\frac{2}{n(n+1)}$, of being picked first, if the first one picked is specified as being purple. Thus, the probability that the first marble came from the bag with a given number, k, of purples is $\frac{2k}{n(n+1)}$. After the first one is picked, there are $k - 1$ purples among the $n - 1$ marbles left in the bag, and the probability of the second pick being purple is $\frac{k-1}{n-1}$. The probability that the first purple marble came from that bag and that the second marble was also purple is the product of the individual probabilities:

$$\frac{2k(k-1)}{n(n+1)(n-1)}.$$

The overall probability for all of the possible combinations is the sum of these values for all values of k from 0 to n:

$$\sum_{k=0}^{n} \frac{2k(k-1)}{n(n+1)(n-1)}$$

$$= \frac{2}{n(n+1)(n-1)} \sum_{k=0}^{n}(k^2 - k)$$

$$= \frac{2}{n(n+1)(n-1)} \left(\frac{n(n+1)(2n+1)}{6} - \frac{n(n+1)}{2} \right)$$

$$= \frac{2}{n-1} \left(\frac{2n+1}{6} - \frac{1}{2} \right)$$

$$= \frac{2}{n-1} \left(\frac{2n+1-3}{6} \right)$$

$$= \frac{4}{6} = \frac{2}{3}.$$

(319) *Problem.*   A box contains p white balls and q black balls, and beside the box lies a large pile of black balls. Two balls chosen at random (with equal likelihood) are taken out of the box. If they are of the same colour, a black ball from the pile is put into the box; otherwise, the white ball is put back into the box. The procedure is repeated until the last two balls are removed from the box and one last ball is put in. What is the probability that this last ball is white?

Solution.   The key to solving this problem revolves around the fact that the number of white balls in the box will never change parity. That is, if it is odd it stays odd and if it is even it stays even, since the only way a white ball is removed is if it is joined in its removal by another white ball. Therefore, when we come down to one last ball remaining in the box, it will be white if we started with an odd number of white balls and black if we started with an even number of white balls.

Thus, the probability of it being white is 0 if p is even, and 1 if p is odd.

(320) *Problem.*   A box contains 20 balls numbered $1, 2, \ldots, 20$. If three balls are randomly taken out of the box, without replacement, what is the probability that the number on one of the balls will be the average of the other two?

Solution. There are several ways to approach this problem. We will only consider one of them here.

If we are to satisfy the condition, then the largest and smallest numbers must be both even or both odd, in which case there is only one value that will work for the number in the middle. The number of ways to choose two even numbers from the given set is $\binom{10}{2} = 45$ and the number of ways to choose two odd numbers from the given set is also $\binom{10}{2} = 45$.

This provides us with a total of 90 acceptable ways to choose three numbers from the given set so that one of these numbers is the average of the other two. Now the total number of ways to choose the three numbers is $\binom{20}{3} = 1140$. Thus, the probability is the ratio $90/1140 = 3/38 = 0.078947$.

(321) *Problem.* A test on blood samples infected with hepatitis has the probability (p) of detecting the disease. In one very large batch of mostly healthy samples, the test found 196 hepatitis infections. After these 196 samples were removed, the same test was run again, and 84 more infected samples were discovered (p obviously is not close to 1). How accurate is the test for hepatitis? That is, find p. How many more infected samples remain after the second test?

Solution. Let h be the number of infected blood samples after the second test. Then there were in total $h + 196 + 84 = h + 280$ infected blood samples initially and $h + 84$ following the first test. Since there were 196 infected samples found in the first test, we have

$$(h + 280)p = 196,$$

and since there were 84 infected samples found in the second test, we also get

$$(h + 84)p = 84.$$

By substituting the second equation in the first we get $84 + 196p = 196$ which yields $p = 4/7$, from which we can solve for $h = 63$. Thus, the probability of detecting the hepatitis on a single test is $4/7$ and there were 63 undetected infected blood samples remaining after the second test.

(322) *Problem.* "Take you on in billiards," said Cue to Ball.

"Sure," Ball said, "I've got time for five games."

"I ought to beat you," said Cue.

"Yeah;" said Ball, "you know, the probability that you win three games is the same as the probability that you win four games."

What is the probability that Cue wins all five games?

Solution. Let p be the probability that Cue wins a given game. Then the probability that Ball wins a given game is $1 - p$. In a set of five games the probability that Cue wins exactly 3 of them is $p^3(1-p)^2$ times the number of ways to select which 3 of the 5 games Cue actually wins, that is, $\binom{5}{3}p^3(1-p)^2$. Similarly the probability that Cue wins exactly 4 of the five games is $\binom{5}{4}p^4(1-p)$. Setting these equal we get

$$\frac{5!}{3!\,2!}p^3(1-p)^2 = \frac{5!}{4!\,1!}p^4(1-p),$$
$$10p^3(1-p)^2 = 5p^4(1-p),$$
$$2(1-p) = p \qquad \text{if } p \neq 0 \text{ and } p \neq 1,$$
$$p = \frac{2}{3}.$$

The case $p = 0$ is ruled out since Cue states he ought to beat Ball. The case $p = 1$ cannot be ruled out. Thus, the probability that Cue wins all five games is either 1 or

$$\binom{5}{5}p^5(1-p)^0 = \left(\frac{2}{3}\right)^5 = \frac{32}{243} \approx 0.13.$$

(323) *Problem.* An urn contains w white balls ($w \geq 3$) and r red balls. If 3 balls were to be withdrawn without replacement, the probability they would all be white is p. An extra white ball in the urn would increase this probability by a third of its value. If r is as great as these conditions allow, how many red balls are there in the urn?

Solution. There are $\binom{w+r}{3}$ ways of selecting 3 balls from the urn without replacement. Of these there are $\binom{w}{3}$ ways of selecting 3 white balls. Thus,

$$p = \frac{\binom{w}{3}}{\binom{w+r}{3}} = \frac{\dfrac{w!}{3!(w-3)!}}{\dfrac{(w+r)!}{3!(w+r-3)!}}. \tag{1}$$

Similarly, with an extra white ball, we have

$$\frac{4}{3}p = \frac{\binom{w+1}{3}}{\binom{w+r+1}{3}} = \frac{\dfrac{(w+1)!}{3!(w-2)!}}{\dfrac{(w+r+1)!}{3!(w+r-2)!}}. \tag{2}$$

Substituting (1) in (2) we get

$$\frac{4}{3} \cdot \frac{\dfrac{w!}{3!(w-3)!}}{\dfrac{(w+r)!}{3!(w+r-3)!}} = \frac{\dfrac{(w+1)!}{3!(w-2)!}}{\dfrac{(w+r+1)!}{3!(w+r-2)!}},$$

$$\frac{4}{3} = \frac{\dfrac{w+1}{w-2}}{\dfrac{w+r+1}{w+r-2}} = \frac{(w+1)(w+r-2)}{(w-2)(w+r+1)},$$

$$4(w-2)(w+r+1) = 3(w+1)(w+r-2),$$

$$r(w-11) = -w^2 + w + 2,$$

$$r = \frac{-w^2 + w + 2}{w - 11} = -w - 10 - \frac{108}{w - 11}$$

$$= -w - 10 + \frac{108}{11 - w}.$$

Since w is a positive integer ≥ 3, in order to keep r finite and positive, we must have $w < 11$. Since r needs to be an integer, we also require that $11 - w$ divides evenly into 108; this eliminates 3, 4, and 6 as possible values for w.

Since we want r as large as possible, we need to make $108/(11-w)$ as large as possible, which clearly occurs if $11 - w$ is as small as possible, which means that w itself should be as large as possible, namely $w = 10$. In this case, we get $r = -10 - 10 + 108 = 88$.

Thus, there are 88 red balls in the urn (and 10 white ones) if we are to have as many red balls as possible and still satisfy all the conditions.

(324) *Problem.* Persons $p_1, p_2, \ldots, p_n$, $n > 1$, have been assigned seats $s_1, s_2, \ldots, s_n$, respectively, on a tour bus.

The passengers board in numerical order. Unable to follow directions, passenger p_1 chooses a seat (uniformly) at random.

Subsequent passengers (upon boarding) take their assigned seat if available, or choose a vacant seat at random if their assigned seat is not available. Find the probability that p_n sits in her assigned seat s_n.

Solution. Let P_n be the probability that passenger p_n will sit in seat s_n. We will use mathematical induction to show that $p_n = \frac{1}{2}$ for $n > 1$. First, with $n = 2$, if passenger p_1 chooses seat s_1, then passenger p_2 has probability 1 of getting the correct seat. Otherwise, p_2 has probability 0 of getting the correct seat. Thus, $P_2 = \frac{1}{2}(1 + 0) = \frac{1}{2}$. Thus, our claim is true for $n = 2$.

Now, suppose $n > 2$ and that $P_m = \frac{1}{2}$ for $1 < m < n$. If passenger p_1 chooses seat s_1, then (as above) passenger p_n has probability 1 of sitting in seat s_n, and if p_1 chooses seat s_n, then passenger p_n has probability 0 of sitting in s_n. If passenger p_1 chooses seat s_k where $1 < k < n$, then passenger p_i sits in seat s_i for $1 < i < k$, and passenger p_k will have to choose from seat s_1 or seats s_j for $k + 1 \leq j \leq n$.

Since this is essentially the same situation as if passenger p_k were p_1 with $n - k + 1$ passengers and seats, we conclude from the induction hypothesis that the probability that passenger p_n sits in seat s_n is $\frac{1}{2}$ for each of the $n - 2$ choices of k, $1 < k < n$. Therefore,

$$P_n = \left(\frac{1 + 0 + (n - 2)\frac{1}{2}}{n} \right) = \frac{1}{2},$$

which completes the induction.

(325) *Problem.* A three-person jury has two members each of whom independently has probability p of making the correct decision and a third member who flips a coin for each decision. A one-person jury has probability p of making the correct decision.

Which jury has the better probability of making the correct decision if majority rules in the three-person jury?

Solution. The two conscientious jurors in the three-person jury agree on the correct decision with a probability of p^2. They will disagree with each other with a probability of $2p(p-1)$, since either one could be correct and the other incorrect; in this case when they consult the flippant juror, the probability of a correct decision is halved to $p(p-1)$.

Hence, the overall probability of a correct decision is $p^2 + p(p-1) = p$, the same as that of a one-person jury.

(326) *Problem.* 35 persons per thousand have high blood pressure. 80% of those with high blood pressure drink, and 60% of those without high blood pressure drink. What percentage of drinkers have high blood pressure?

Solution. Let us assume a population base of 1000 persons. Of the 35 who have high blood pressure 80% or 28 drink; of the 965 who do not have high blood pressure 60% or 579 drink. Thus, there are $28 + 579 = 607$ drinkers in the population. Of these 28 have high blood pressure. Therefore, the percentage is:

$$\frac{28 \times 100}{607} \approx 4.613\%.$$

Chapter 9

Triangle Mathematics

(327) *Problem.* Determine x in the equilateral triangle shown below:

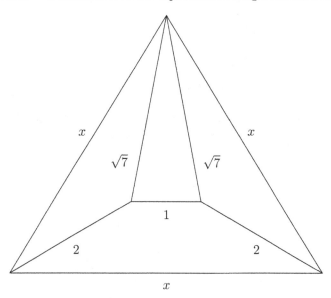

Solution. Let us first name the vertices and draw some altitudes. See diagram on next page.

In any equilateral triangle the ratio of the altitude to the base is $\sqrt{3}/2$. Thus, the altitude of $\triangle ABC$ is $\frac{x\sqrt{3}}{2}$. Now $DF = \frac{1}{2}$. Applying the Theorem of Pythagoras to $\triangle ADF$, we have

$$AF^2 = AD^2 - DF^2 = 7 - \frac{1}{4} = \frac{27}{4}.$$

Therefore, $AF = 3\sqrt{3}/2$.

277

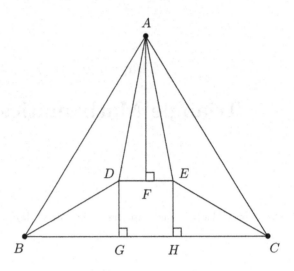

Then

$$EH = DG = \frac{x\sqrt{3}}{2} - \frac{3\sqrt{3}}{2} = \frac{\sqrt{3}}{2}(x-3).$$

Since $BG = HC$, we have $HC = (x-1)/2$. Thus, by applying the Theorem of Pythagoras to $\triangle ECH$, we have

$$EH^2 + HC^2 = EC^2$$

$$\left(\frac{\sqrt{3}}{2}(x-3)\right)^2 + \left(\frac{x-1}{2}\right)^2 = 2^2$$

$$\frac{3(x^2-6x+9)}{4} + \frac{x^2-2x+1}{4} = 4$$

$$3x^2 - 18x + 27 + x^2 - 2x + 1 = 16$$

$$4x^2 - 20x + 12 = 0$$

$$x^2 - 5x + 3 = 0,$$

from which we conclude that

$$x = \frac{5 \pm \sqrt{13}}{2}.$$

Since $(5 - \sqrt{13})/2 < 1$, we may discard that possibility, and we are left with

$$x = \frac{5 + \sqrt{13}}{2}.$$

(328) *Problem.* Let ABC be a right-angled triangle of area 1. Let A', B', and C' be the points obtained by reflecting A, B, C, respectively, in their opposite sides. Find the area of triangle $A'B'C'$.

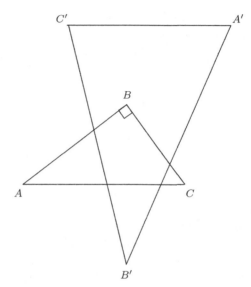

Solution. Let angle B be the right angle. Let BB' meet AC in D and $A'C'$ in D'.

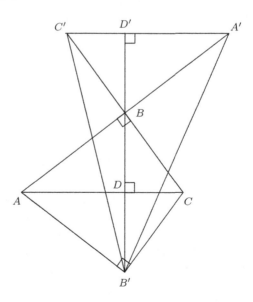

Clearly BB' is perpendicular to AC, since AC is the line of reflection of B into B'. Since triangles ABC and $A'BC'$ are congruent and since B is a right angle, lines AC and $A'C'$ are parallel. This implies that BB' extended is also perpendicular to $A'C'$. Thus $B'D'$ is the altitude to the base $A'C'$ in the triangle $A'B'C'$.

Since triangle $AB'C$ is also congruent to the above two triangles, we see that $B'D$, DB, and BD' are all the same length, which is the altitude to the base AC in triangle ABC. Thus triangle $A'B'C'$ has the same length base as triangle ABC and altitude three times as long. Thus the area is three times as great, making the area of triangle $A'B'C'$ equal to 3.

(329) *Problem.*   If the sides of a triangle are 4, 5, and 6, prove that the largest angle is exactly double the smallest angle.

Solution. First of all, it should be clear that the largest angle is opposite the largest side and the smallest angle is opposite the smallest side. Let ABC be the triangle, with angle A the smallest angle and angle B the largest angle. Let α be the measure of angle A and let β be the measure of angle B. Construct the perpendicular altitude from C to the line AB meeting it at D. Let h be the length of this altitude, and let x be the length of BD. By using the Theorem of Pythagoras in triangles BCD and ACD, we obtain:

$$h^2 = 4^2 - x^2 \qquad \text{and} \qquad h^2 = 6^2 - (5 - x)^2.$$

By eliminating h in the above and solving for x, we get $x = \frac{1}{2}$. Now construct the point E on AB such that triangle ECD is congruent to triangle BCD. Thus, CE has length 4, DE has length $x = \frac{1}{2}$, and angle CED has measure β. However, that implies that AE has length $5 - 2x = 4$, from which it follows that triangle ACE is isosceles, and angle ACE has measure α. Since angle CED is the exterior angle of triangle ACE its measure is the same as the sum of the measures of the two opposite interior angles, namely A and ACE. That is, $\beta = 2\alpha$, which is what we set out to prove.

(330) *Problem.* ADB and AEC are isosceles right triangles, right-angled at D and E, respectively, described outside $\triangle ABC$. F is the mid-point of BC. Prove that DFE is an isosceles right-angled triangle. (Diagram on next page.)

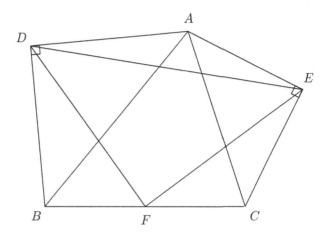

Solution. Let G and H be the mid-points of AB and AC, respectively. It is easy to see that $DG = AG = HF$ and that $HE = AH = GF$.

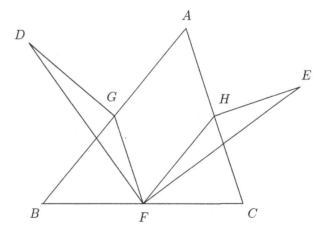

Also, $\angle DGF = \angle FHE$ since the extensions of their corresponding sides are mutually perpendicular. Hence triangles DGF and FHE are congruent with $DF = FE$. Furthermore, since corresponding sides of triangles DGF and FHE are mutually perpendicular, DF is perpendicular to FE. Thus DFE is a right-angled isosceles triangle.

(331) *Problem.*   How many triangles have the form shown below, where n is a positive integer, and x is a real number, $0 < x \le 1$?

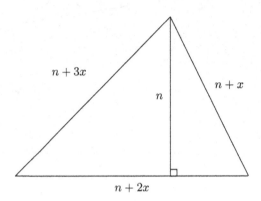

Solution.   The area of a triangle with sides a, b, and c is given by Heron's formula:

$$A = \sqrt{s(s-a)(s-b)(s-c)},$$

where $s = (a + b + c)/2$. Using the substitutions $a = n + x$, $b = n + 2x$, and $c = n + 3x$, and setting Heron's formula equal to the more traditional formula for the area of a triangle, we have

$$\sqrt{\frac{3n+6x}{2} \cdot \frac{n+4x}{2} \cdot \frac{n+2x}{2} \cdot \frac{n}{2}} = \frac{n(n+2x)}{2}$$

$$\frac{3(n+2x)(n+4x)(n+2x)n}{16} = \frac{n^2(n+2x)^2}{4},$$

$$\frac{3(n+4x)}{4} = n,$$

$$12x = n.$$

Since $0 < x \le 1$ and n is a positive integer, there are exactly 12 triangles of the required type, namely $(x, n) = (\frac{1}{12}, 1), (\frac{2}{12}, 2), \dots, (1, 12)$.

(332) *Problem.*   A circle of radius 1 cm is inscribed in an equilateral triangle.

A smaller circle is inscribed at each vertex, tangent to the circle and to two sides of the triangle.

The process is continued with progressively smaller circles. What is the sum of the circumferences of all the circles?

Solution. First use some elementary geometry to show that the equilateral triangle has altitude 3 cm (exercise!).

Including the initial circle the sum of the diameters of the circles at each vertex is simply the altitude of the triangle namely 3 cm. Since the initial circle has diameter 2 cm, the sum of the diameters of the circles at each vertex excluding the initial circle is 1 cm. Thus, the sum of the diameters of all the circles is 5 cm, implying that the sum of the circumferences of all the circles is 5π cm.

(333) *Problem.* A circle with radius 5 is escribed to triangle ABC, being tangent to side BC and also to sides AB and AC extended. The centre of the circle is 13 units from vertex A of the triangle. What is the perimeter of the triangle?

Solution. Let O be the centre of the circle, and let D, E, and F be the points of tangency of the circle with AB (extended), BC, and AC (extended), respectively. Then OD is perpendicular to AD, and by the Theorem of Pythagoras we have

$$(AD)^2 = (AO)^2 - (OD)^2$$
$$= 13^2 - 5^2$$
$$= 144,$$

from which we see that $AD = 12$. Similarly $AF = 12$.

Now $BD = BE$ since the tangents from a point are equal in length. Also $CE = CF$. If we let p be the perimeter of triangle ABC, then we see that

$$p = AB + BC + AC$$
$$= AB + (BE + EC) + AC$$
$$= (AB + BE) + (AC + CE)$$
$$= (AB + BD) + (AC + CF)$$
$$= AD + AE = 24.$$

(334) *Problem.* AD is a median of a triangle ABC and ℓ is a line passing through A. E is a point on ℓ such that CE is parallel to AD. F and G are the feet of the perpendiculars from E and B, respectively, on AD.

Prove that EG is parallel to BF.

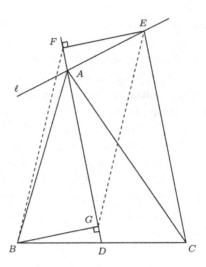

Solution.   First extend the line BG to meet CE at H.

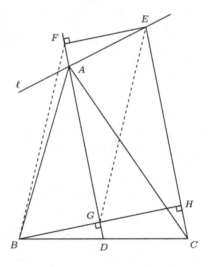

Since CE is parallel to AD and $BG \perp AD$, we see that $BH \perp CE$. Thus, $EFGH$ is a rectangle, which means that AF is parallel to BH. Next, we note that $BG = GH$ (since $BD = CD$ and CE is parallel to AD). Since BG and AF are equal in length and parallel to each other, we conclude that $EFBG$ is a parallelogram. It immediately follows that GE is parallel to BF.

(335) *Problem.* Farmer Brown has a rectangular plot of prime land to divide between his two sons. He divided the rectangle $ACEF$ into 4 parts as shown in the figure (the areas of 3 of the parts are shown in the figure). He gave one son the triangle ABF, and the other the triangle FDE. Then he gave them this challenge: The son who first determines the area of the triangle FDB gets that piece of land plus the triangle BCD.

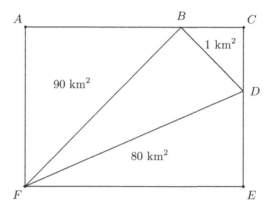

What is the area of the triangle FDB?

Solution. Let the dimensions of the rectangular plot be $x \times y$, where x and y are both measured in km.

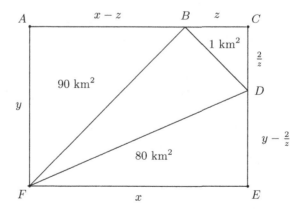

Also, let the length of BC be denoted by z km as shown in the diagram above. Since the area of $\triangle BCD$ is 1 km^2, we conclude that CD has length $2/z$ km. Computing the areas of the triangles

ABF and *DEF* we have:

$$y(x - z) = 180$$

$$x\left(y - \frac{2}{z}\right) = 160$$

or $\quad xy - zy = 180$

$$xy - \frac{2x}{z} = 160.$$

Eliminating xy above and rearranging we get $yz^2 + 20z - 2x = 0$, whose only positive solution is

$$z = \frac{\sqrt{100 + 2xy} - 10}{y}.$$

Substituting this in the first equation above yields:

$$xy - \sqrt{100 + 2xy} + 10 = 180,$$
$$xy - 170 = \sqrt{100 + 2xy},$$
$$(xy)^2 - 340xy + 170^2 = 100 + 2xy,$$
$$(xy)^2 - 342xy + 170^2 - 10^2 = 0,$$

which is a quadratic in xy. Solving for xy we have:

$$xy = \frac{342 \pm \sqrt{342^2 - 4(170^2 - 10^2)}}{2}$$
$$= 171 \pm \sqrt{171^2 - (170^2 - 10^2)}$$
$$= 171 \pm \sqrt{441} = 171 \pm 21,$$

which means that $xy = 150$ or 192. But xy represents the area of the original rectangular plot which contains more than $90 + 80 + 1 = 171$ km^2. That is, $xy > 171$, which implies that $xy = 192$.

When we subtract from this the areas of the 3 triangles whose areas are given we are left with the area of $\triangle BDF = 192 - 90 - 80 - 1 = 21$ km^2.

(336) *Problem.* In $\triangle ABC$, M is a point on BC such that $BM = 5$ and $MC = 6$. If $AM = 3$ and $AB = 7$, determine the exact value of AC.

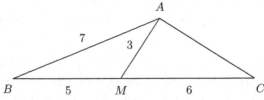

Solution. Applying the Law of Cosines to $\triangle AMB$, we have

$$\cos \angle AMB = \frac{5^2 + 3^2 - 7^2}{2 \cdot 5 \cdot 3} = -\frac{1}{2}$$

whence $\angle AMB = 120°$, implying that $\angle AMC = 60°$.
Applying the Law of Cosines to $\triangle AMC$, we get

$$AC^2 = 3^2 + 6^2 - 2 \cdot 3 \cdot 6 \cdot \cos \angle AMC = 9 + 36 - 35 \cdot \tfrac{1}{2} = 27,$$

from which it follows that $AC = 3\sqrt{3}$.

Alternate Approach: Drop a perpendicular from A to BC at P.
Let $x = MP$, $h = AP$, and $y = AC$ as in the diagram below:

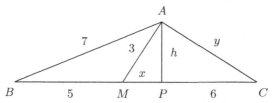

Now, $\angle AMB$ is obtuse since $7^2 > 3^2 + 5^2$, so that P is located
between M and C as in the diagram. Thus, we have $h^2 = 3^2 - x^2$
from $\triangle APM$. In $\triangle APB$. we have

$$7^2 = (5 + x)^2 + h^2 = 25 + 10x + x^2 + 9 - x^2 = 34 + 10x,$$

which yields $x = 1.5$. From $\triangle APC$, we obtain:

$$y^2 = (6 - x)^2 + h^2 = 36 - 12x + x^2 + 9 - x^2 = 45 - 12x = 27.$$

Therefore, $AC = 3\sqrt{3}$.

(337) *Problem.* The inscribed circle of a right triangle ABC is tangent
to the hypotenuse AB at D. If $AD = x$ and $DB = y$, find the area
of the triangle in terms of x and y.

Solution. First of all, let us recall that if tangents are drawn from
a point P external to a circle $\mathcal{C}$, then the points of tangency are
equi-distant from P. Thus, if we let E, F be the points of tangency
of the inscribed circle with the sides AC, BC, respectively, then we
have $AE = x$ and $BF = y$ (see the diagram on the next page). Let
I be the centre of the inscribed circle. Then $IECF$ is a square of
side a, which means that $CE = CF = a$, the radius of the inscribed
circle.

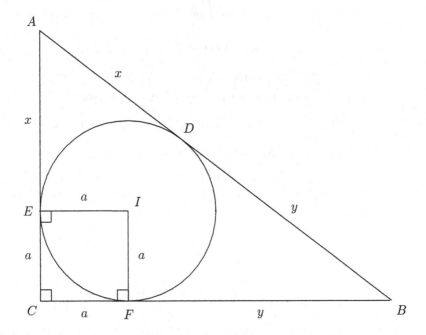

The Pythagorean Theorem yields

$$(x + a)^2 + (y + a)^2 = (x + y)^2,$$
$$x^2 + 2ax + a^2 + y^2 + 2ay + a^2 = x^2 + 2xy + y^2,$$
$$2a(x + y) + 2a^2 = 2xy,$$
$$a(x + y + a) = xy.$$

Thus, the area of the triangle is

$$\frac{(x + a)(y + a)}{2} = \frac{xy + ax + ay + a^2}{2}$$
$$= \frac{xy + a(x + y + a)}{2}$$
$$= \frac{xy + xy}{2}$$
$$= xy.$$

(338) *Problem.*   Shipwrecked on an island which is in the shape of an
equilateral triangle, a sailor builds a hut so that the total of its
distances to the three sides of the triangle is a minimum. Where is
the best place on the island to locate the hut?

Solution.   Let ABC be equilateral triangle of side length k. Let H
be the proposed location of the hut. Let h_a, h_b, h_c be the distances
from H to the sides BC, CA, AB, respectively. Now the area of
$\triangle ABC$ is the sum of the areas of $\triangle HBC$, $\triangle HCA$, and $\triangle HAB$
which is

$$\frac{kh_a}{2} + \frac{kh_b}{2} + \frac{kh_c}{2} = \frac{k}{2}(h_a + h_b + h_c).$$

Since this area is constant, as is the value k, we conclude that
$h_a + h_b + h_c$ is also a constant. Hence, any point on the island is
as good as any other for the location of the hut.

(339) *Problem.*   Let ABC be a right triangle with right angle at A. Let
the inscribed circle of $\triangle ABC$ touch the side BC at point D. Prove
that the area of $\triangle ABC$ is equal to the product of the lengths of
BD and CD.

Solution.   As usual, we denote the lengths of BC, CA, and AB
by a, b, and c, respectively. Let the inscribed circle also touch AC
at E and AB at F as in the diagram below.

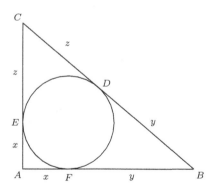

The segments AE and AF have the same length since both are
tangents to the circle from A. Let $x = AE = AF$. Similarly, let
$y = BF = BD$ and $z = CE = CD$. If we denote the semiperimeter
of the triangle by s, then $s = \frac{1}{2}(a + b + c) = x + y + z$. Clearly,
$BD = y = s - (x + z) = s - b$ and $CD = z = s - (x + y) = s - c$.
Now

$$BD \cdot CD = (s - b)(s - c) = s^2 - s(b + c) + bc$$

$$= \left(\frac{1}{4}(a + b + c)^2 - \frac{1}{2}(a + b + c)(b + c) + bc \right)$$

$$= \frac{1}{4}((a + b + c)^2 - 2(a + b + c)(b + c) + 4bc)$$

$$= \frac{1}{4}((a + b + c)a - (a + b + c)(b + c) + 4bc)$$

$$= \frac{1}{4}(a^2 - (b + c)^2 + 4bc) = a^2 - b^2 - c^2 + 2bc$$

$$= \frac{1}{4}(2bc) \qquad \text{since } a^2 = b^2 + c^2$$

$$= \frac{1}{2}bc,$$

which is the area of $\triangle ABC$.

(340) *Problem.*   Consider the isosceles triangle ABC in the diagram below, which has $\angle BAC = 20°$. On AC, one of the equal sides, a point D is marked off so that $AD = BC = a$. Find the measure of $\angle ABD$.

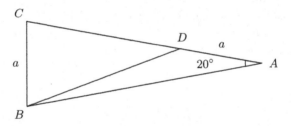

Solution.   First we have a "Proof Without Words", which shows that the angle we seek has measure $10°$.

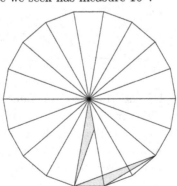

A second method involves constructing an equilateral triangle ABE on the exterior of $\triangle ABC$, as shown in the diagram below.

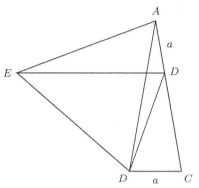

Then
$$\angle EAD = 60° + 20° = 80° = \angle ACB.$$
Since we also have $EA = AB = AC$ and $AD = BC$, we see that $\triangle EAD$ is congruent to $\triangle ACB$. Therefore, $ED = EA = EB$, which implies that $\triangle EBD$ is isosceles. Since $\angle BED = 60° - 20° = 40°$, we have $\angle EBD = 70°$; whence
$$\angle ABD = \angle EBD - \angle EBA = 70° - 60° = 10°.$$

(341) *Problem.* A square $PQRS$ is inscribed in a semicircle with centre O and diameter AB such that P and Q lie on the diameter AB (see the diagram below). With C lying on the minor arc RB, a (right-angled) triangle ABC having the same area as $PQRS$ is inscribed in the semicircle. Show the incentre I of $\triangle ABC$ (that is, the centre of the circle inscribed in $\triangle ABC$) is located at the intersection of RQ and SB, and that
$$\frac{RI}{IQ} = \frac{SI}{IB} = \frac{1 + \sqrt{5}}{2},$$
the so-called "Golden Ratio".

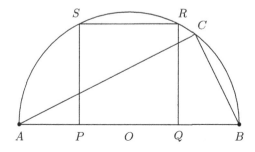

Solution. Let I be the intersection of RQ and SB. Since $\triangle ROS$ and $\triangle BOC$ have the same area (they both have one half the area of the square $PQRS$), and since they are both isosceles triangles with the radius of the semicircle as the length of the equal sides, we must have $\overset{\frown}{RS} = \overset{\frown}{BC}$, and

$$\overset{\frown}{CS} = \overset{\frown}{BR} = \overset{\frown}{SA}.$$

Since $\overset{\frown}{AS} = \overset{\frown}{SC}$, we see that $\angle ABS = \angle SBC$. Thus, BS bisects $\angle ABC$. Drop a perpendicular from I to BC meeting BC at M.

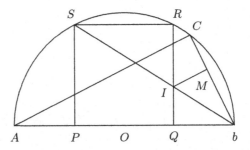

If we denote the radius of the semicircle by ρ, then

$$SP = \frac{2\rho}{\sqrt{5}}, \qquad QB = \rho\left(1 - \frac{1}{\sqrt{5}}\right), \qquad PB = \rho\left(1 + \frac{1}{\sqrt{5}}\right).$$

We also have

$$MC = BC - BM = RS - QB = SP - QB = \frac{\rho(3 - \sqrt{5})}{\sqrt{5}},$$

and, from similar triangles, we also get

$$IM = IQ = QB\left(\frac{SP}{PB}\right) = \frac{\rho(3 - \sqrt{5})}{\sqrt{5}}.$$

Therefore, $IM = MC$, which means that $\triangle IMC$ is an isosceles right-angled triangle. Hence, IC bisects $\angle ACB$, which makes I the incentre of $\triangle ABC$, since it is the intersection of two of the angle bisectors of $\triangle ABC$.

Finally, from similar triangles, we see that

$$\frac{RI}{IQ} = \frac{SI}{IB} = \frac{RS}{QB} = \frac{SP}{QB} = \frac{1 + \sqrt{5}}{2}.$$

(342) *Problem.*   Prove that there is no triangle whose altitudes are of length 4, 7, and 10 units.

Solution.   Suppose there is such a triangle. We will show this leads to a contradiction. Let the sides perpendicular to the altitudes of 4, 7, and 10 be denoted a, b, and c, respectively. Since the area of the triangle can be computed as one half the product of any side times its corresponding altitude we see that $4a = 7b = 10c$. Now examine the following:

$$70(b + c) = 70b + 70c = 10 \times 7b + 7 \times 10c = 10 \times 4a + 7 \times 4a$$
$$= 40a + 28a = 68a < 70a.$$

This implies that $a > b + c$, but it is impossible in any triangle for one side to exceed the sum of the lengths of the other two sides.

(343) *Problem.*   In triangle ABC, the lines BD and BE are trisectors of angle B, while CD and CE are trisectors of angle C. Suppose that point E is closer to BC than point D. Prove that angles BDE and EDC are equal.

Solution.   Simply consider triangle DBC. The line BE bisects $\angle CBD$ and the line CE bisects $\angle BCD$. Since the three angle bisectors of any triangle meet at a common point, the bisector of $\angle BDC$ must also pass through the point E. This means that DE bisects angle BDC, and we are done.

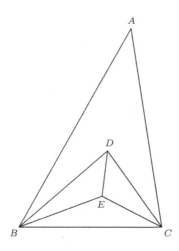

(344) *Problem.*   A 1-acre field in the shape of a right triangle has a post at the mid-point of each side. See diagram below.

A sheep is tethered to each of the side posts and a goat to the post on the hypotenuse. The ropes are just long enough to let each animal reach the two adjacent vertices. What is the total area the two sheep have to themselves, that is, the area the goat cannot reach?

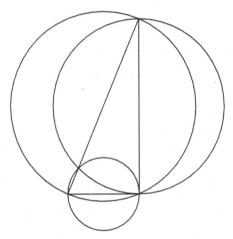

Solution.   This problem only makes sense if we assume that the animals can graze outside of the 1 acre field, since the goat can clearly graze over the entire field (and then some). With this in mind we let ABC represent the triangle with AB the hypotenuse. The first observation that we need to make is that the goat (tethered to the mid-point of the hypotenuse) can just reach the third vertex C of the triangle. Let us consider the semicircle on AB which passes through the point C, together with the two semicircles on the sides AC and BC which enclose the areas that the sheep can graze, but cannot be grazed by the goat. Let us denote the areas of these semicircles by G, S_1 and S_2, respectively. If we further let T represent the area of the original triangle and X the area we wish to find, then we have

$$G + X = S_1 + S_2 + T,$$
$$\text{or} \quad X = (S_1 + S_2 - G) + T.$$

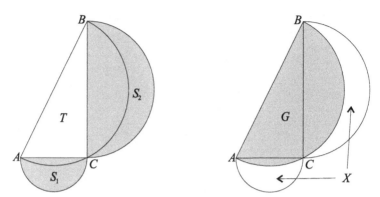

Since S_1, S_2 and G are proportional to the squares of the two sides and of the hypotenuse, we deduce that $S_1 + S_2 - G = 0$ by the Theorem of Pythagoras. Hence, $X = T$. That is, the area that the sheep can graze, but which cannot be grazed by the goat is the same as the area of the triangle, that is, 1 acre.

Chapter 10

Miscellaneous

(345) *Problem.* A biologist wants to calculate the number of fish in a lake. On May 1, she catches a random sample of 60 fish, tags them, and releases them. On September 1, she catches a random sample of 70 fish and finds that 3 of them are tagged.

To calculate the number of fish in the lake on May 1, she assumes that 25% of these fish are no longer in the lake on September 1 (because of death and emigrations), that 40% of the fish present on September 1 were not in the lake on May 1 (because of births and immigrations) and that the number of untagged fish and tagged fish in the September 1 sample are representative of the total population.

What does the biologist calculate for the number of fish in the lake on May 1?

Solution. Let x be the fish population in the lake on May 1, and let y be the fish population in the lake on September 1. Then on September 1 the number of fish from May 1 that are still in the lake is $\frac{3}{4}x$, and they represent $\frac{3}{5}y$ with respect to the population size on September 1. Thus, $y = \frac{5}{4}x$. (That is, the fish population has increased by 25% in the 4 month period.) Since $\frac{3}{4}$ of the fish from May 1 are still in the lake on September 1, the same can be said for the tagged fish in the lake on May 1 (which numbered 60). Thus, on September 1 there are still 45 tagged fish in the lake. Therefore, the ratio of tagged fish to the total fish population must be the same in the sample as in the lake:

$$\frac{3}{70} = \frac{45}{y} = \frac{45}{\frac{5}{4}x},$$

from which we can conclude that $x = 840$.

(346) *Problem.*   Depicted below is a rectangle partitioned into 9 squares. If the small square has sides of length 1, what are the lengths of the sides of the rectangle?

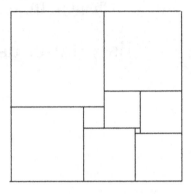

Solution.   Consider the squares labelled as in the diagram below. If we let x be the length of the side of square E, then squares F, H, G, and B have sides of length $x+1$, $x+2$, $x+3$, and $2x+1$, respectively. Now by considering the line separating squares E and G, we conclude that square D has sides of length 4. We can now conclude that squares C and A have sides of length $x+7$ and $x+11$, respectively. By considering the line separating squares A and B, we have

$$(x+11) + 4 = (2x+1) + x,$$
$$\text{that is;} \quad x+15 = 3x+1$$
$$\text{or} \quad x = 7.$$

This completely solves the rectangle, which must have dimensions 33 across and 32 down.

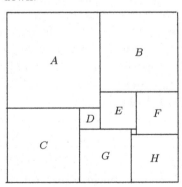

(347) *Problem.* 101 marbles are numbered from 1 to 101. The marbles are divided over two baskets A and B. The marble numbered 40 is in basket A. This marble is removed from basket A and put in basket B. The average of all the numbers on the marbles in A increases by $\frac{1}{4}$. The average of all the numbers of the marbles in B increases by $\frac{1}{4}$. How many marbles were there originally in basket A?

Solution. Let n_A and n_B be the initial number of marbles in the baskets A and B, respectively. Then we have $n_B = 101 - n_A$. Let a and b be the initial totals of all the marbles in the baskets A and B, respectively.

Then $a + b = 101(102)/2 = 5151$, or $b = 5151 - a$. The information on the average values leads to the following equations:
$$\frac{a - 40}{n_A - 1} - \frac{a}{n_A} = \frac{1}{4},$$
$$\frac{b + 40}{n_B + 1} - \frac{b}{n_B} = \frac{1}{3}.$$
Upon removing fractions in the above and simplifying we get
$$n_A^2 + 159n_A - 4a = 0,$$
$$n_B^2 - 159n_B + 4b = 0.$$
Using $N_B = 101 - n_A$ and $b = 5151 - a$, we see that the second of the above equations can be expressed as
$$n_A{}^2 - 43n_A - 4a + 14746 = 0.$$
We now have two equations relating n_A and a. If we subtract one from the other we get $202n_A - 14746 = 0$ which reduces to $n_A = 73$. Thus, there are initially 73 marbles in basket A (and 28 marbles in basket B).

Note: The interested reader can find several ways to divide the marbles with 73 in basket A (including the one numbered 40) and 28 in basket B such that the requirements are met. One such method is to place all the marbles from 22 to 94 in basket A and the rest in basket B.

(348) *Problem.* A rocket car accelerates from 0 km per hour to 240 km per hour in a test run of one kilometre. If the acceleration is not allowed to increase (but it may decrease) during the run, what is the longest time the run can take?

Solution. Consider the graph of the velocity $v(t)$ (t in hours, v in km per hour) of the car's motion (see the graph on the next page, where T is the time the car takes to travel the kilometre).

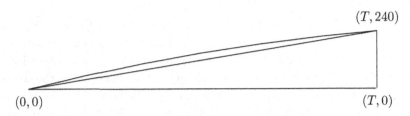

$(T, 240)$

$(0,0)$ $(T, 0)$

The tangent line at any point has slope equal to the acceleration at that point and, because the latter does not increase, the slope of the tangent line does not increase, that is, the graph of $v(t)$ is concave down (or at least is never concave up).

The area under the curve is equal to the distance traveled (namely 1 kilometre), and is not less than the area of the triangle with vertices $(0,0)$, $(T,0)$, $(T,240)$. Hence, the area of the triangle is $\frac{1}{2}(240)T < 1 =$ the area under the curve, which implies that $T \leq \frac{1}{120}$ hours, that is, the time T is at most 30 seconds.

(349) *Problem.* A carpenter had four pieces of wood cut in the shape of an isosceles trapezoid as indicated in the figure below. The x and y in each case are a whole number of inches. Although the x and y are different in each piece, the area is the same for all four pieces and is indeed a whole number of square inches less than 30. What are the dimensions of the four pieces?

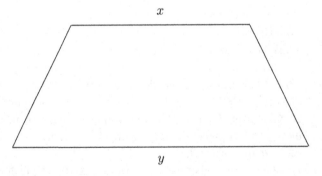

x

y

Solution. Although the area A can be calculated, in terms of x and y, directly from the given diagram, the easiest way of finding A is to note that four identical pieces of this form can be put together to form a hollow square or a square frame as in the figure on the next page. (Indeed, this was what the carpenter used the pieces for.)

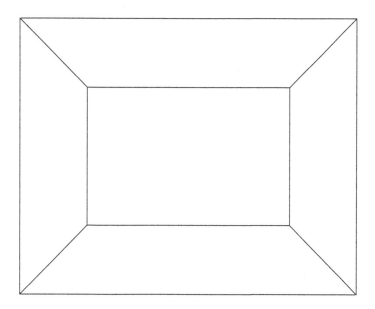

Thus,

$$A = \frac{y^2 - x^2}{4} = \left(\frac{y+x}{2}\right)\left(\frac{y-x}{2}\right).$$

If we set $m = \frac{y+x}{2}$ and $n = \frac{y-x}{2}$, then $m+n = y$ and $m-n = x$, and also $A = mn$. From this and from the conditions of the problem it is evident that A must be a number which can be factored into two integers in at least four ways. There is only one such integer less than 30, namely 24.

Thus, $24 = 1 \times 24 = 2 \times 12 = 3 \times 8 = 4 \times 6$, which yield the corresponding values for x and y of:

$$(25, 23), (14, 10), (11, 5), (10, 2).$$

(350) *Problem.* A sequence $\{a_n\}$ of real numbers is defined by

$$a_1 = 1, \qquad a_{n+1} = 1 + a_1 a_2 \cdots a_n \quad \text{for } n \geq 1.$$

Show that

$$\sum_{n=1}^{\infty} \frac{1}{a_n} = 2.$$

Solution. Let the partial sum S_n be defined as

$$S_n = \sum_{k=1}^{n} \frac{1}{a_k}.$$

We wish to show that $\lim_{n \to \infty} S_n = 2$. Consider the following empirical observations:

n	a_n	$1/a_n$	S_n
1	1	1	1
2	2	$\frac{1}{2}$	$\frac{3}{2}$
3	3	$\frac{1}{3}$	$\frac{11}{6}$
4	7	$\frac{1}{7}$	$\frac{83}{42}$
5	43	$\frac{1}{43}$	$\frac{3611}{1806}$

Observe that for $1 \leq n \leq 4$ we have

$$S_n = 2 - \frac{1}{a_{n+1} - 1}.$$

Let us now prove this is true for all $n \geq 1$ by mathematical induction.

We can clearly observe that it is true for $n = 1$, 2, 3, and 4. Now suppose that it is true for some $n = k \geq 4$. That is, for some $k \geq 4$, we have

$$S_k = 2 - \frac{1}{a_{k+1} - 1}.$$

We must show that it is true for $n = k + 1$. But

$$S_{k+1} = S_k + \frac{1}{a_{k+1}} = 2 - \frac{1}{a_{k+1} - 1} + \frac{1}{a_{k+1}}$$

$$= 2 - \left(\frac{1}{a_{k+1} - 1} - \frac{1}{a_{k+1}} \right) = 2 - \left(\frac{a_{k+1} - (a_{k+1} - 1)}{a_{k+1}(a_{k+1} - 1)} \right)$$

$$= 2 - \frac{1}{a_{k+1}(a_{k+1} - 1)} = 2 - \frac{1}{a_{k+1}(a_1 a_2 \cdots a_k)}$$

$$= 2 - \frac{1}{a_{k+2} - 1},$$

which proves the induction step. Thus, $S_n = 2 - 1/(a_{n+1} - 1)$ for all $n \geq 1$. Since $\lim_{n \to \infty} a_n = \infty$, we have $\lim_{n \to \infty} 1/a_n = 0$. Therefore, $\lim_{n \to \infty} S_n = 2$, which was to be shown.

(351) *Problem.*   A man enter a store and chooses a $10 hat. Since he does not carry enough money on him, he makes the following proposition to the merchant:

"If you lend me as much money as I have in my pocket, I will buy this $10 hat."

The merchant accepts and our gentleman buys the hat. He then goes to another store and repeats the operation, this time in order to buy a $10 pair of sandals. In a third store, this same procedure allows him to buy a $10 umbrella. After this last purchase, he has no more money. If we assume that the man lived in Alberta (when there was no sales tax of any kind!), how much money did the man have in his pocket when he entered the first store?

Solution.   Let the man have x dollars when he entered the first store. After purchasing the $10 hat, he has in his pocket $2x - 10$ dollars. After purchasing the $10 pair of sandals, he has in his pocket $2(2x - 10) - 10 = 4x - 30$ dollars. Finally, after purchasing the $10 umbrella, he has $2(4x - 30) - 10 = 8x - 70$ dollars. But we are told that he had no money left after this purchase. Therefore,

$$8x = 70,$$
$$\text{so that} \quad x = 8.75.$$

Thus, he started with $8.75 in his pocket.

(352) *Problem.*   For $480 a man bought a number of equally-priced cows, each cow costing an integral number of dollars. After the death of three of the cows, the man disposed of the remaining cows, selling all of them for identical integral dollar amounts, and making a profit of $15 on the entire transaction.

If the price the man paid for each cow was $1 less than the price for which he sold each cow, find the number of cows the man originally bought.

Solution.   Let x be the number of cows the man originally bought. Then he paid $480/x$ dollars for each cow. Since the selling price was $1 more than the buying price, the selling price for each cow was $(480/x) + 1$ dollars.

The number of cows sold was $x - 3$, since three cows died before he could sell them. Since he made a profit of $15 on the transaction, he must have received $495 in total at the sale. Thus, another way of calculating the selling price per cow is $495/(x - 3)$.

Therefore,

$$\frac{480}{x} + 1 = \frac{495}{x-3},$$

or $480(x-3) + x(x-3) = 495x.$

This yields:

$$x^2 - 15x - 1440 = 0.$$

From the quadratic formula we see that $x = -30$ or 48. Since he cannot have purchased a negative number of cows, he must have purchased 48 cows.

(353) *Problem.*   Find the sum of

$$1 \cdot 1! + 2 \cdot 2! + 3 \cdot 3! + \cdots + n \cdot n!$$

where $n! = n \cdot (n-1) \cdot (n-2) \cdots 2 \cdot 1.$

Solution.   Notice that the summation in question would be a little easier to add if each of the summands of the form $i \cdot i!$ was replaced by $(i+1) \cdot i!$, which is seen to simply be $(i+1)!$. Let $S(n)$ be the sum in question. Now let

$$T(n) = 1! + 2! + 3! + \cdots + n!.$$

Then clearly $S(n) = S(n) + T(n) - T(n)$. However, by combining the obvious terms one can see that $S(n) + T(n) = T(n+1) - 1$. Thus, we conclude that

$$S(n) = T(n+1) - 1 - T(n).$$

But since $T(n+1) - T(n)$ equals $(n+1)!$, we further conclude

$$S(n) = (n+1)! - 1.$$

(354) *Problem.*   The circumference of a rear wheel of a wagon is one foot more than the circumference of a front wheel. If a front wheel makes twenty-two more revolutions than a rear wheel in travelling a mile, find the radius in feet of the smaller wheel.

Solution.   Let x be the circumference (in feet) of the smaller wheel. (Although we are asked to find the radius of this wheel, the algebra is simpler if we first solve for the circumference and then compute the radius from this). Then the circumference of the rear (larger) wheel is $x + 1$ feet. Let n be the number of revolutions of the front (smaller) wheel in travelling a mile (5280 feet). Then the number of revolutions of the rear wheel is $n - 22$. Since both wheels travel the same distance (one mile) we must have

$$nx = 5280 \qquad \text{and} \qquad (n - 22)(x + 1) = 5280.$$

The second equation becomes

$$nx - 22x + n - 22 = 5280,$$

from which we can conclude that $n = 22(x+1)$, since first equation states that $nx = 5280$. By using this value for n in the first equation we obtain the quadratic equation

$$x^2 + x - 240 = 0,$$

which has as its roots 15 and -16. Since the circumference of a wheel cannot be negative, we conclude that it must be 15 feet. Consequently, the radius is $\frac{15}{2\pi}$ feet.

(355) *Problem.* Each of 5 men has a whole number of dollars. They decide to share their dollars. The first man gives each of the other men an amount equal to the amount that each has at that time. The second man then does the same thing and so do the third, fourth and fifth men. At the end of the exchange it turns out that each of the men has the same number of dollars. What is the minimum amount of money that each man had at the beginning?
Solution. At the very end each man had the same amount of money, say n dollars. Thus, the total amount of money being handled was $5n$ dollars. Just prior to this, the fifth man had given each man as many dollars as he currently had. Therefore, before the fifth man's handout, the others had $\frac{1}{2}n$ dollars each and the fifth man had $3n$ dollars. Similarly, before the fourth man's handout, the first three men had $\frac{1}{4}n$ dollars each, the fifth man had $\frac{3}{2}n$ dollars and the fourth man had $\frac{11}{4}n$ dollars; before the third man's handout, the first two men had $\frac{1}{8}n$ dollars each, the fourth man had $\frac{11}{8}n$ dollars, the fifth man had $\frac{3}{4}n$ dollars and the third man had $\frac{21}{8}n$; before the second man's handout, the first man had $\frac{1}{16}n$ dollars, the third man had $\frac{21}{16}n$ dollars, the fourth man had $\frac{11}{16}n$ dollars, the fifth man had $\frac{3}{8}n$ dollars and the second man had $\frac{41}{16}n$ dollars. Finally, before the first man's handout, the second man had $\frac{41}{32}n$ dollars, the third man had $\frac{21}{32}n$ dollars, the fourth man had $\frac{11}{32}n$, the fifth man had $\frac{3}{16}n$ dollars and the first man had $\frac{81}{32}n$. Clearly, if n is positive and an integer, it must be a multiple of 32. It is equally clear that any multiple of 32 yields a solution. The

smallest solution set occurs when n is $32. This gives values of $81, $41, $21, $11 and $6, respectively, for the five men.

(356) *Problem.*   A train passes through a long tunnel in the mountains. The train is 125 metres long, and travels at 40 kilometres per hour. Just as the last car is completely inside the tunnel, a man begins walking from the end of the train to the front. He walks at 5 kilometres per hour and reaches the front of the train just as it emerges into daylight. How long is the tunnel?

Solution.   Let t be the time taken for the man to walk the entire length of the train. This can be calculated to be 0.025 hr. During this time the train moves 1 km.

When we started timing the man on his walk, the train was already 125 m into the tunnel. From our calculations we see that the train moves yet another 1 km = 1000 m before it leaves the tunnel. Thus the length of the tunnel is 1125 m, or 1.125 km.

(357) *Problem.*   The circumference of a "roulette Wheel" is divided into 36 sectors to which the numbers 1, 2, 3, ..., 36 are assigned in some arbitrary manner. Show that there are three consecutive sectors with the property that the sum of their assigned sectors is at least 56.

Solution.   The proof will be by contradiction. Suppose that there is no set of three consecutive sectors with the property that the sum of their assigned values is at least 56. Worded another way: every set of three consecutive sectors has assigned values whose sum at is most 55.

Consider a partition of the sectors into 12 groups of three consecutive sectors. The sum of the assigned values for each of the 12 sectors is at most 55 for a total of at most 660. On the other hand, every number from 1 to 36 inclusive must appear exactly once in this partition.

Consequently, the sum must be exactly
$$1 + 2 + 3 + \cdots + 35 + 36 = 666,$$
which is obviously impossible.

(358) *Problem.*   A train travelling from Aytown to Beetown meets with an accident after 1 hour. It is stopped for $\frac{1}{2}$ hour, after which it proceeds at four-fifths of its usual rate, arriving at Beetown 2 hours late. If the train had covered 80 miles more before the accident, it would have been just 1 hour late. What is the usual rate of the train?

Solution. There are a number of ways to express all the detail of this problem into mathematical equations. However, if we pare away all the unnecessary information we conclude that the time taken to drive the extra 80 miles at $\frac{4}{5}$ of the usual rate is one hour longer than the time taken to drive it at the usual rate.

Therefore, if we let v be the usual rate of the train in miles per hour, and if we recall that time is distance/rate, we see that

$$\frac{80}{(4/5)v} = \frac{80}{v} + 1$$

(both sides of which are expressions for time). The solution to this equation is $v = 20$ mph.

(359) *Problem.* A car travelling in a northerly direction carries a small flag fixed to a vertical pole. There is a strong and steady wind blowing from the West. When the car is travelling at 40 km/hr, the flag makes an angle of 60° with respect to a North-South line. At what speed must the car travel for the flag to make an angle of 30° with respect to the North-South line?

Solution. This is a vector problem. The wind can be represented by a vector. The northerly movement of the car induces a southerly force on the car, which can also be represented by a vector. The direction of the flag is given by the resultant of these two vectors. Let w be the speed of the wind. Then $w = 40 \tan 60°$ km/hr. Now, let s be the speed of the car needed for the flag to make the angle 30° with the North-South line. Then $s = w \tan 60° = 120$ km/hr.

(360) *Problem.* Start with an infinite supply of 7 cent stamps and 10 cent stamps. How many different postage values can <u>not</u> be made with these stamps? For example, there is no way of getting 13 cents using just 7 cent and 10 cent stamps, while $1.44 can be formed with thirteen 10 cent stamps and two 7 cent stamps.

Solution. Since one of the denominations is 10 cents, the units digit of any postage value is affected only by the number of 7 cent stamps. For example, to get a digit of 1 in units position we require three 7 cent stamps (21 cents). This means that any postage value ending in a 1 can be represented except 1 cent and 11 cents. If we use a similar argument for each other possible value of the units digit of the postage value, we get a total of 27 different values which cannot be made up, namely 1, 2, 3, 4, 5, 6, 8, 9, 11, 12, 13, 15, 16, 18, 19, 22, 23, 25, 26, 29, 32, 33, 36, 39, 43, 46, and 53.

For an interesting variation which requires a different solution technique, suppose that the values of the stamps are 7 cents and 12 cents. In general, if m and n have a greatest common divisor of 1, then the largest value that cannot be provided with stamps valued at m cents and n cents is given by $mn - (m+n)$. Note here that $7 \times 10 - (7 + 10) = 53$, the largest unattainable value found above.

(361) *Problem.*   The mathematician Augustus De Morgan, who lived in the nineteenth century, when asked the year of his birth, countered with: "I was x years old in the year x^2." When was he born?

Solution.   Clearly, x must be an integer whose square lies between 1800 and 1899. The only such integer is the number 43. Since the square of 43 is 1849, De Morgan was 43 in 1849. Thus, he was born in 1806.

(362) *Problem.*   A man travels m feet due north at 2 minutes per mile. He returns due south to his starting point at 2 miles per minute. What is the average speed in miles per hour for the entire trip?

Solution.   In order to find the average speed we need to determine the total distance traveled and the total time elapsed. Let x be the distance m measured in miles. Clearly the total distance traveled is $2x$ miles. To find the time elapsed we will determine the time elapsed on each of the two legs separately. Going north the time elapsed was $2x$ minutes, while heading south $x/2$ minutes passed by (use the relation $d = rt$, and solve for t!). Thus, the total time elapsed was $2.5x$.

The average speed then is $2x/2.5x$, which simplifies to 0.8 miles per minute or 48 miles per hour.

(363) *Problem.*   In a factory, square tables of size 1 m × 1 m are tiled with four tiles each of size 50×50 cm^2. All tiles are the same, and decorated in the same way with an asymmetric pattern such as the letter "J". How many different types of tables can be produced in this way?

Solution.   Since the tiles cannot be flipped, there are four possible orientations. Number the corners of each tile with the numbers 1, 2, 3, and 4:

For example,

Now, for each table there is a 4-tuple consisting of the numbers of the tile corners at each of the corners of the table listed counterclockwise starting at any corner.

For example,

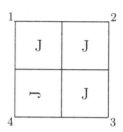

gives $(1, 2, 3, 1)$ or $(1, 1, 2, 3)$, etc. Two tables are the same just in the case one can be rotated to match the other, e.g. $(1, 1, 2, 3)$, $(1, 2, 3, 1)$, $(2, 3, 1, 1)$, $(3, 1, 1, 2)$ all give the same table.

Most configurations, like the one above, admit 4 different labellings; in these, the first pair of numbers can be chosen in 16 different ways, and the second pair in 15 ways (since the second pair must be distinct from the first pair), giving $(16 \times 15)/4 = 60$ different tables. (We have divided by 4 since there are 4 different labellings for each such table.) Some other configurations have only two labellings, e.g. $(1, 3, 1, 3)$ and $(3, 1, 3, 1)$; in these, the first pair can be chosen in $\binom{4}{2} = 6$ ways (since the two numbers must be different) and the second pair is then determined, 6 more tables. Finally, there are the configurations which have only one labelling, namely $(1, 1, 1, 1)$, $(2, 2, 2, 2)$, $(3, 3, 3, 3)$, and $(4, 4, 4, 4)$. This gives a grand total of $60 + 6 + 4 = 70$ tables.

(364) *Problem.* Among a group of people are a number of dogs and octopuses. In total there are 18 heads and 64 legs counted. If there are twice as many people as dogs, how many octopuses are there?
Solution. Let p, d, o be the numbers of people, dogs and octopuses, respectively. Then we have

$$p + d + o = 18$$

if we count heads

and $2p + 4d + 8o = 64$

if we count legs.

But $p = 2d$. Therefore, $8d + 8o = 64$, from which we can conclude that $d + o = 8$, whence $p = 10$, which implies that $d = 5$ and the number of octopuses is 3.

(365) *Problem.*   A two-pan balance is inaccurate since its balance arms are of different lengths and its pans are of different weights. Three objects of different weights A, B, and C are each weighed separately.

When placed on the left hand pan, they are balanced by weights A_1, B_1, C_1, respectively.

When A and B are placed on the right hand pan, they are balanced by A_2 and B_2, respectively.

If $A_1 = 25$ gm, $B_1 = 9$ gm, $C_1 = 1$ gm, $A_2 = 43$ gm, $B_2 = 18$ gm, determine the true weight of C.

As an extra challenge determine the true weight of C in general in terms of A_1, B_1, C_1, A_2, B_2.

Solution.   We will illustrate the general solution and then apply it to the above case with the values given.

Let W_1, W_2 be the weights of the left and right hand pans, respectively. Let L_1, L_2 be the lengths of the left and right hand balance arms, respectively. Then the five weighings produce the following results:

$$(A + W_1)L_1 = (A_1 + W_2)L_2, \qquad (1)$$
$$(B + W_1)L_1 = (B_1 + W_2)L_2, \qquad (2)$$
$$(C + W_1)L_1 = (C_1 + W_2)L_2, \qquad (3)$$
$$(A_2 + W_1)L_1 = (A + W_2)L_2, \qquad (4)$$
$$(B_2 + W_1)L_1 = (B + W_2)L_2. \qquad (5)$$

By considering equations (1) and (4) and eliminating A, we get

$$A_2 L_1^2 - A_1 L_2^2 = (W_2 L_2 - W_1 L_1)(L_2 + L_1).$$

Similarly, from equations (2) and (5) we get

$$B_2 L_1^2 - B_1 L_2^2 = (W_2 L_2 - W_1 L_1)(L_2 + L_1).$$

From these two equations, we see that

$$\frac{L_2}{L_1} = \sqrt{\frac{A_2 - B_2}{A_1 - B_1}}.$$

If we let r be this value then we can solve equation (3) to obtain:
$$C = \frac{A_2 - A_1 r^2}{1 + r} + C_1 r.$$
Thus, for the values given, $C = 3$ gm.

(366) *Problem.* When Maharaja Ram Singh died, he left 3465 gold pieces to be divided equally among his children. Each wife had the same number of children and this number was 8 less than the number of wives per harem, which in turn was 4 more than the number of harems and 4 less than the number of gold pieces that each child received. How many children did Ram Singh have?

Solution. Let the number of children per wife be x. Then the number of wives per harem is $x + 8$, the number of harems is $x + 4$ and the number of gold pieces that each child receives is $x + 12$. Then
$$x(x + 4)(x + 8)(x + 12) = 3465.$$
If we let $y = x + 6$ this translates into
$$(y - 6)(y - 2)(y + 2)(y + 6) = 3465,$$
$$\text{that is,} \quad (y^2 - 4)(y^2 - 36) = 3465,$$
$$\text{or} \quad y^4 - 40y^2 - 3321 = 0,$$
which factors into
$$(y^2 - 81)(y^2 + 41) = 0,$$
from which we get $y^2 = 81$ or $y = \pm 9$. Since $x = y - 6$, we must have $y = 9$ and $x = 3$ and the number of children is $3465/(3 + 12) = 231$.

(367) *Problem.* A coin collector had a table with an exactly circular hole in it, where long ago had been an inkwell. He had two pure gold coins of the same thickness; the larger coin exactly fitted the hole, and the smaller one, when slid gradually over the hole, tipped into it when its edge reached the centre of the hole. The larger coin weighed 6 oz.; what was the weight of the smaller coin?

Solution. When the small coin reaches the point where it tips, its diameter AD must be a chord of the circular hole. Let C be the centre of the hole and let B be the centre of the coin. Then $CB \perp AD$ because AC and DC are radii of the hole. If we let the radius of the coin be 1, we see that $AB = BD = 1$. Then $AC = \sqrt{2}$, and since the coins have the same thickness, their weights are proportional to their areas, which in turn are proportional to the squares of their respective radii. Hence the big coin weighs twice as much as the smaller one, since $(\sqrt{2})^2 = 2$. Therefore, the smaller coin weighs 3 oz.

(368) *Problem.* Tickets for the senior prom were $10.00 for boys and $6.50 for girls. Although there were more boys than girls at the dance, the percentage of boys who did not go was twice the percentage of girls who did not go. Knowing this percentage and the total senior class enrollment, one can deduce the total receipts for the affair. If this enrollment is between 60 and 100, what was the total attendance at the prom?

Solution. Let n be the number of students in the senior class, of which b were boys and let p be the proportion of girls who did not go to the senior prom. Then the number of girls was $n - b$ and the proportion of boys who did not go to the prom was $2p$. Therefore, the total receipts R (in dollars) is

$$R = 10b(1 - 2p) + 6.5(n - b)(1 - p)$$
$$= 3.5b - 13.5bp + 6.5n - 6.5np.$$

Since we are told that knowledge of n and p is sufficient to deduce the total receipts, the above expression must not involve the term b, that is, the coefficient of b must be 0.

Therefore, $3.5 = 13.5p$ which yields $p = \frac{7}{27}$. This means that the proportion of boys who did not go to the prom is $2p = \frac{14}{27}$.

Thus, the number of girls in the senior class is a multiple of 27 and the number of boys is a multiple of 54. Since the total is between 60 and 100, we see that the number of girls is 27 and the number of boys is 54. Then there were 26 boys and 20 girls in attendance at the prom for a total of 46 students.

(369) *Problem.* The hands of an accurate clock have lengths 3 and 4. Find the distance between the tips of the hands when that distance is increasing most rapidly.

Solution. Solution using calculus: First note that the minute hand and hour hand of the clock move in a clockwise direction at the rate of 1 revolution and $\frac{1}{12}$ revolution per hour, respectively.

Because we are dealing with calculus, we must convert this to radian measure – that is, they move at a rate of 2π radians and $\frac{\pi}{6}$ radians per hour, respectively.

This implies that the angle θ between them decreases at the rate of $2\pi - \frac{\pi}{6}$ radians per hour. That is,

$$\frac{d\theta}{dt} = -2\pi - \frac{\pi}{6} = -\frac{11\pi}{6}.$$

Now from the Law of Cosines, if we let x be the distance between the tips of the two hands, we have $x^2 = 25 - 24\cos\theta$. Differentiating implicitly with respect to time t, we obtain

$$2x\frac{dx}{dt} = 24\sin\theta\frac{d\theta}{dt},$$

or $\qquad \dfrac{dx}{dt} = \dfrac{12\sin\theta}{x}\cdot\dfrac{-11\pi}{6} = -22\pi\dfrac{\sin\theta}{x}.$

It is this last quantity, the rate of change of the distance with respect to time, that we must maximize. We do that by taking its derivative and setting it to 0. Let $v = \frac{dx}{dt}$. Then

$$\frac{dv}{dt} = -22\pi\left(\frac{x\cos\theta\dfrac{d\theta}{dt} - \sin\theta\dfrac{dx}{dt}}{x^2}\right)$$

$$= -22\pi\left(\frac{\dfrac{-11\pi}{6}x\cos\theta - \sin\theta\left(-22\pi\dfrac{\sin\theta}{x}\right)}{x^2}\right)$$

$$= \frac{121\pi^2}{3x^2}\left(x\cos\theta - \frac{12\sin^2\theta}{x}\right).$$

Then, the only way that this expression can be 0 is if

$$x^2\cos\theta = 12\sin^2\theta = 12 - 12\cos^2\theta.$$

Now, from our original observation about the Law of Cosines, we have $\cos\theta = \frac{25-x^2}{24}$.

When we substitute for $\cos\theta$ and simplify, we get $x^4 = 49$, from which we deduce that $x = \sqrt{7}$.

Solution avoiding calculus: The movement of the tip of each hand is given by a vector perpendicular to the hand. Since the hands turn at fixed rates, the angle between them also turns at a fixed rate. We may associate all the movement with one of the hands relative to the other hand, in particular with the hour hand relative to the minute hand. The movement of the tip of the hour hand is a vector of fixed length (because the rotation is at a fixed rate), and is perpendicular to the hour hand.

It is clear that the distance between the tips of the two hands will be increasing most rapidly when the component of this vector at the end of the hour hand in the direction of the line joining the tips is maximum.

But this obviously occurs when the vector is in line with the line joining the tips; that is, when the hour hand is perpendicular

to the line joining the tips of the two hands. By the Theorem
of Pythagoras, we conclude that the distance at that moment is
$\sqrt{16-9} = \sqrt{7}$.

(370) *Problem.*   "You wait out here," Mary told her husband as she
turned back to re-enter the big store. "I won't be long." Steve was
only too glad to miss another battle with the hordes of Christmas
shoppers who thronged in the aisles inside. "But won't you need
some more cash?" he asked.

Mary checked in her bag. "I've got plenty, dear," she assured him.
"It's only one little thing I forgot. I spent a quarter of my money
in there already, but I've still got five dollars more than I would
have if I'd spent a third instead of a quarter."

That gave Steve something to think about while he waited. How
much would you say she still had in her bag?

Solution. Let x be the number of dollars Mary had to begin with.
The amount she has now is $\frac{3}{4}x$. On the other hand it is also $5 + \frac{2}{3}x$.
Setting these two expressions equal to each other we obtain the
value of $x = 60$, which means that Mary had \$60 to begin, and
now has \$45 still left.

(371) *Problem.*   A 4 ft. wide flag is to be designed with a red square
between two equal vertical rectangles of white. (Of course, this
means that the height of the square determines the height of the
flag.) How large should the red square be in order to give the largest
amount of white on the sides? This question can be answered
without calculus; can you find the answer without calculus?

Solution. Whether you place the red square in the middle of the
flag or on one end makes no difference to the problem. Let us place
it on the left side. (See diagram below.) The 45° diagonal is the
locus of its free corner, and the large rectangle to the right, with
height h and width w, is the total area of white left.

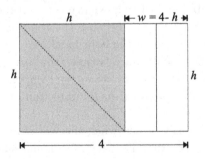

When $h = w = 2$, then the area of white equals 4. Since $h + w = 4$, if one of them is $2 + x$, the other is $2 - x$ and the area of white is $4 - x^2$, which is less than 4 if $x \neq 0$. Therefore, the maximum white area is achieved when $h = 2$.

(372) *Problem.* Alan can run around a circular track in 40 seconds. Betty, running in the opposite direction, meets Alan every 15 seconds. How many seconds does it take Betty to run around the track?

Solution. Let t be Betty's time (in seconds) to run around the track. Together Alan and Betty cover the length of the entire track every 15 seconds. If we denote by l the length of the track, then Alan's speed is $l/40$ and Betty's speed is l/t. Thus they separate at the rate of $(l/40) + (l/t)$. Consequently, we have

$$\left(\frac{l}{40} + \frac{l}{t} \right) \times 15 = l$$
$$\text{or} \quad (t + 40)15 = 40t,$$
$$\text{therefore} \quad t = 24.$$

(373) *Problem.* Tom, Dick and Harry started out on a 100-mile journey. Tom and Harry went by tandem bicycle at the rate of 25 mph, while Dick walked at the rate of 5 mph. After a certain distance, Harry got off and walked on at 5 mph, while Tom went back for Dick and got him to the destination at the same time that Harry arrived. How many hours were required for the trip? (Assume that the bicycle always travels at the same speed of 25 mph.)

Solution. Let t_1 be the time when Tom and Harry were on the bicycle; let t_2 be the time when Tom was alone on the bicycle; and let t_3 be the time when Tom and Dick were on the bicycle. Then by considering the distances that each of Tom, Dick and Harry covered we get the following system:

$$25t_1 - 25t_2 + 25t_3 = 100$$
$$5t_1 + 5t_2 + 25t_3 = 100$$
$$25t_1 + 5t_2 + 5t_3 = 100.$$

This system can be solved for the unique solution $t_1 = 3$, $t_2 = 2$, and $t_3 = 3$. Thus, the entire trip requires 8 hours.

(374) *Problem.* Side AB of a square $ABCD$ is divided into n segments in such a way that the sum of the lengths of the even-numbered segments equals the sum of the lengths of the odd-numbered segments.

Lines parallel to AD are drawn through each point of division, and each of the n "strips" thus formed is divided by the diagonal BD into a left region and a right region. Show that the sum of the areas of the left regions with odd numbers is equal to the sum of the areas of the right regions with even numbers.

Solution. Let x be the sum of the areas of the left regions with odd numbers; let y be the sum of the areas of the left regions with even numbers. Let z and w be the corresponding values for the right regions. We must show that $x = w$. Since the diagonal BD divides the area of the square in half, we must have

$$x + y = z + w.$$

But we are also given

$$y + w = x + z.$$

Subtracting one from the other we obtain

$$x - w = w - x,$$

from which it is clear that $x = w$, and we are through.

(375) *Problem.*   The integers $1, 2, \ldots, n$ are placed in order so that each value is either strictly bigger than all preceding values or is strictly smaller than all preceding values.

In how many ways can this be done?

Solution.   Consider beginning this sequence at the end. The last number in the sequence must be either 1 or n, that is, one of two possibilities.

This means that the first $n - 1$ positions must consist either of the integers $1, 2, \ldots, n - 1$ or $2, 3, \ldots, n$.

A similar argument can be used for each of the last terms of this shorter sequence.

Thus, there are two choices for each of the numbers in the sequence taken from last to first, excepting the first position, where there is only one value remaining.

Thus, there are 2^{n-1} possible sequences.

(376) *Problem.*   Alan and Bruce run a race from opposite ends of a straight track. One minute after they pass, Alan finishes. Another three minutes passed before Bruce finished. How much faster was Alan than Bruce?

Solution. Let v_A and v_B be the respective speeds of Alan and Bruce during the race. Let t be the elapsed time when they pass. If d is the length of the track, we have

$$d = (v_A + v_B)t. \qquad (*)$$

Now the total time elapsed for Alan to run the race is $t+1$ minutes, and for Bruce the time elapsed is $t + 4$. From this we obtain

$$v_A(t + 1) = d$$
$$v_B(t + 4) = d.$$

Using the value of d from (*) and simplifying we obtain the system:

$$v_A = v_B t$$
$$4v_B = v_A t,$$

from which we obtain $v_A^2 = 4v_B^2$, which means that Alan runs twice as fast as Bruce.

(377) *Problem.* Two proofreaders were asked to read a certain manuscript. Adams found 200 errors and Baker found 220 errors, 175 of which were also found by Adams. How many mistakes did they both miss (approximately)? Can you generalize this?

Solution. We will show the solution to the general problem and use it to answer the particular question asked. Suppose there were m mistakes altogether in the manuscript. Suppose that Adams discovered a of them, Baker discovered b of them, and that there were c mistakes discovered by both.

Now Adams has an efficiency rating of a/m, since he discovered a of the m mistakes. On the other hand we have a second estimate of his efficiency, namely he discovered c of the b mistakes discovered by Baker. We will assume that Adams and Baker were consistent in their work. This leads to the equation:

$$\frac{a}{m} = \frac{c}{b}$$

from which we get

$$m = \frac{ab}{c},$$

and the number of mistakes missed by both Adams and Baker is

$$m - a - b + c = \frac{ab}{c} - a - b + c = \frac{ab - ac - bc + c^2}{c} = \frac{(a - c)(b - c)}{c}$$

which can be computed from the known quantities.

For our example, $a = 200$, $b = 220$ and $c = 175$. This yields a value of:

$$\frac{(a-c)(b-c)}{c} = \frac{25 \cdot 45}{175} = 6.43,$$

giving an estimate of 6 or 7 mistakes missed by both.

(378) *Problem.*   Ann spends from 3:00 pm to 9:00 pm walking along a level road, up a hill, and home again. On the level road she walks 4 km/hr, uphill 3 km/hr, and downhill 6 km/hr. Find the distance that Ann walked.

Solution.   Let d_1 be the distance on the level and let d_2 be the distance on the hill. Then the time taken to walk on the level, uphill, downhill, and returning on the level are:

$$\frac{d_1}{4}, \quad \frac{d_2}{3}, \quad \frac{d_2}{6}, \quad \frac{d_1}{4},$$

respectively (all measured in hours). The sum of these times, which is $(d_1 + d_2)/2$, must be 6 hours. Thus, the total distance walked, which is $2(d_1 + d_2)$, must be 24 km.

(379) *Problem.*   The interior angles of a convex polygon, measured in degrees, form an arithmetic sequence. The smallest angle is 120° and the common difference is 5°. Find the number of sides of all such polygons.

Solution.   Let n be the number of sides of the polygon. Then, the sum of the interior angles is $(n-2)180°$. (To see this simply join one vertex to each of the others to form $n - 2$ disjoint triangles; all the interior angles of the polygon are captured in the interior angles of the triangles and vice versa; since the sum of the interior angles of a triangle is 180°, the rest follows.) But the sum of the interior angles of the polygon must also be the sum of the arithmetic sequence, namely $n(120°) + (\frac{1}{2})n(n-1)(5°)$. Therefore,

$$120n + \frac{5n(n-1)}{2} = 180n - 360.$$

This equation simplifies to

$$n^2 - 25n + 144 = 0,$$

which factors into

$$(n-9)(n-16) = 0.$$

From this we see that n can be 9 or 16. However, since the polygon is convex, all interior angles must be smaller than 180°. Since these angles start at 120° and grow by 5°, this means that n must be less than 12. Thus, the only solution is $n = 9$.

(380) *Problem.*   A point P in the plane is 6, 7 and 9 units, respectively, from 3 of the 4 vertices of a rectangle in the plane. What is the minimum distance it could be from the fourth vertex? (Can you determine all the possible distances it could be from the fourth vertex?)

Solution.   Let us label the rectangle $ABCD$. Let the distances from P to the vertices A, B, C, and D be a, b, c, and d, respectively. Now draw lines through P parallel to the sides of the rectangle, meeting the (possibly extended) sides AB, BC, CD, and DA at the points E, F, G, and H, respectively (see diagram). Let the lengths AE and EB be denoted by l_1 and l_2; and let the widths BF and FC be denoted by w_1 and w_2. Then by applying the Theorem of Pythagoras several times, we get

$$\begin{aligned}
a^2 + c^2 &= (l_1^2 + w_1^2) + (l_2^2 + w_2^2) \\
&= (l_1^2 + w_2^2) + (l_2^2 + w_1^2) \\
&= d^2 + b^2.
\end{aligned}$$

Now in our problem three of the values a, b, c, and d are given; suppose a, b, and c are the given distances. If d is the smallest possible length, then b must be the largest. Thus, we have $d^2 + 9^2 = 6^2 + 7^2 = 85$, from which we deduce that $a = 2$.

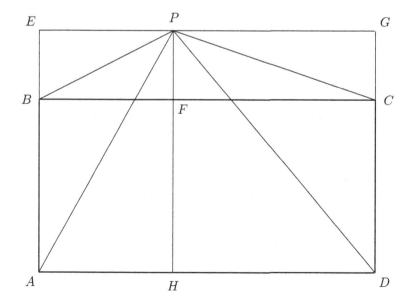

Clearly there are only 3 possible solutions if we drop the minimal condition. They occur by selecting b as 6, 7 or 9. We have already handled the case $b = 9$. It is easy to see that the cases $b = 6$ and $b = 7$ yield $a = \sqrt{94}$ and $a = \sqrt{68}$.

(381) *Problem.* Let $P(x)$ be a polynomial having integer coefficients. Show that if

$$Q(x) = P(x) + 12$$

has at least six distinct integer roots, then $P(x)$ has no integer roots.

Solution. Let r_i be 6 distinct integer roots of $Q(x)$ for $i = 1, \ldots, 6$. The Factor Theorem tells us that

$$Q(x) = q(x)(x - r_1)(x - r_2)(x - r_3)(x - r_4)(x - r_5)(x - r_6).$$

Since

$$Q(x) = P(x) + 12,$$

we have

$$P(x) = q(x)(x - r_1)(x - r_2)(x - r_3)(x - r_4)(x - r_5)(x - r_6) - 12.$$

Now let us suppose that a is an integer which is a root of $P(x)$. Then $P(a) = 0$. But

$$P(a) = q(a)(a - r_1)(a - r_2)(a - r_3)(a - r_4)(a - r_5)(a - r_6) - 12.$$

Therefore, $\prod_{i=1}^{6}(a - r_i)$ divides 12 evenly, and the values $a - r_i$ are 6 distinct integers. One way to find such a set of 6 integers is to keep the absolute values as small as possible: this would mean setting these 6 values to ± 1 and ± 2 first.

The product of these 4 values is already 4, which leaves only one possible remaining integer, namely 3. Thus we can obtain at most 5 such integers, but there must be 6. This contradiction implies that if a is an integer, it cannot be a root of $P(x)$.

(382) *Problem.* It is well known that the equation $y = ax^2 + bx + c$ represents a parabola when $a \neq 0$.

Consider the family of parabolas where a and c are fixed values and b is allowed to vary. Each parabola in this family has a vertex. What does the set of vertices of this family of parabolas look like? Can you express it in terms of a and c?

Solution. Let us go formally through the process of completing the square for a typical member of this family of parabolas (this is the same process that leads to the development of the quadratic formula):

$$y = ax^2 + bx + c = a\left(x^2 + \frac{b}{a}x + \frac{c}{a}\right)$$

$$= a\left(\left(x^2 + \frac{b}{a}x + \frac{b^2}{4a^2}\right) + \frac{c}{a} - \frac{b^2}{4a^2}\right)$$

$$= a\left(\left(x + \frac{b}{2a}\right)^2 + \frac{c}{a} - \frac{b^2}{4a^2}\right)$$

$$= a\left(\left(x - \left(-\frac{b}{2a}\right)\right)^2 + \frac{c}{a} - \frac{b^2}{4a^2}\right)$$

$$= a\left(x - \left(-\frac{b}{2a}\right)\right)^2 + c - \frac{b^2}{4a}.$$

It is clear from this last expression that the vertex of the parabola occurs at the point with coordinates $(-\frac{b}{2a}, c - \frac{b^2}{4a})$. If we set $x = -\frac{b}{2a}$, then the y-coordinate of the family of vertices can be expressed by the equation $y = c - ax^2 = -ax^2 + c$, which is also a parabola. Thus, the vertices of the family of parabolas $y = ax^2 + bx + c$ with a and c fixed determine another parabola, namely $y = -ax^2 + c$.

(383) *Problem.* Five given points in the plane have the following property: of any four of them, three are the vertices of an equilateral triangle. Prove that four of the five points are the vertices of a rhombus with an interior angle equal to 60°.

Solution. Label the points A, B, C, D, and E. Some three of the points, say A, B, C, form an equilateral triangle. If one of the sides of this equilateral triangle, say AB, makes a new equilateral triangle when paired with either D or E, say D, then $ACBD$ is a rhombus with interior angle equal to 60°. Let us suppose instead that neither D nor E is the vertex of an equilateral triangle together with the lines AB, AC, or BC. Now let us consider the three sets of four points $\{D, E, A, B\}$, $\{D, E, A, C\}$, and $\{D, E, B, C\}$. Then two of the three triangles DEA, DEB, and DEC are equilateral. We may assume without any loss of generality that $\triangle DEA$ and $\triangle DEB$ are equilateral. Then $ADBE$ are the vertices of a rhombus with interior angle equal to 60°.

It is also an interesting question to determine the exact configuration that the conditions imply, from which one can deduce the total number of equilateral triangles determined by the five points.

(384) *Problem.*   I have two little bags, of which the contents are identical. Each has in it four blue marbles, four red marbles, and four yellow marbles.

I close my eyes and remove from the first bag enough marbles (but only just enough) to ensure that my selection includes at least one marble of each colour, and at least two of some colour. These marbles I transfer to the second bag.

Now (again closing my eyes) I transfer from the second bag to the first bag enough marbles to ensure that, in the first bag, there will be at least three marbles of each of the three colours.

How many marbles will be left in the second bag?

Solution.   The first operation is simple enough. Five marbles must be transferred, since the first four drawn may all be of the same colour. Now consider the second operation. There are four possibilities (we have designated the three colours by a, b, and c):

Marbles left in first bag				Marbles now in second bag		
a	b	c		a	b	c
0	3	4		8	5	4
1	2	4		7	6	4
1	3	3		7	5	5
2	2	3		6	6	5

In each case it may be necessary to retransfer as many as 12 marbles. For example, in the first case the first nine marbles to be retransferred may all be b and c. And similarly with the other cases. So there will be five marbles left in the second bag.

Alternate solution: In order to guarantee at least 3 of each colour in the first bag at the end of the transfers, it is necessary to have 19 marbles present in the bag (otherwise, one could find 8 of each of two colours and 2 of the third). This leaves only 5 marbles in the second bag.

(385) *Problem.*   Two long distance runners, moving in the same direction around a circular track, each at a constant speed, pass every 12 minutes. If the faster of the two completes a lap in 10 seconds less

than the other, what fraction of the track does the faster runner complete in one second?

Solution. Let the proportion of the circle the faster runner completes in one second be $\frac{1}{x}$ and the proportion for the slower runner be $\frac{1}{y}$. Then the times taken for a full lap are x and y seconds respectively, and so $y - x = 10$. In 12 minutes, that is, 720 seconds, the faster runner completes one extra lap. Thus, $\frac{720}{x} - \frac{720}{y} = 1$ or

$$\frac{720}{x} - \frac{720}{10 + x} = 1,$$
$$7200 + 720x - 720x = x^2 + 10x,$$
$$x^2 + 10x - 7200 = 0,$$
$$(x + 90)(x - 80) = 0.$$

The positive solution to this equation is $x = 80$, so that the faster runner completes $\frac{1}{80}^{\text{th}}$ of the track in one second.

(386) *Problem.* An astronaut lands on the equator of a spherical asteroid. She travels due north 100 km, without reaching the pole, then east 100 km, then south 100 km. She does not pass the same point more than once, and finds that she is due east of her original starting point by 200 km. How many kilometres would she now need to travel by continuing her journey in an easterly direction in order to reach her original starting point?

Solution. Let r be the radius of the asteroid in km. Let θ be the latitude (expressed in radian measure) she reached when she was 100 km north of the equator. Then $r\theta = 100$. Let ϕ be the change in longitude (again measured in radians) as she traveled east. Consider now the circle at the latitude 100 km north of the equator. By looking at any great circle passing through the north pole we see that this circle has radius $r' = r\cos\theta$. Furthermore we must have $r'\phi = 100$, that is, $r\phi\cos\theta = 100$. We are also told that at the equator these two lines of longitude are 200 km apart. Thus, $r\phi = 200$. From these last two results we see that $\cos\theta = \frac{1}{2}$, which implies that $\theta = \frac{\pi}{3}$. From the first equation we have $r = \frac{300}{\pi}$, which says that the equator has circumference $2\pi r = 600$ km. Therefore, there remain

$$600 - 200 = 400 \text{ km}$$

to travel east to return to the starting point.

(387) *Problem.* A, B, C, and D are four different weights. When they are placed on a balance scale, the following observations are made: A and B exactly balance C and D. A and C together outweigh B and D together, C is lighter than D. Arrange the weights in order from heaviest to lightest.

Solution. From the statements we can deduce the following:

$$A + B = C + D \qquad (1)$$
$$A + C > B + D \qquad (2)$$
$$C < D. \qquad (3)$$

Equations (1) and (2) allow us to conclude that

$$A + C > B + D > B + C,$$

which implies that $A > B$. This together with (1) and (2) imply:

$$2B < A + B = C + D < 2D,$$

from which we have $B < D$. But using this together with (1) and (2), we have

$$A + C > B + D > B + A = C + D,$$

which tells us that $A > D$. Since $A + B = C + D$ and $A > D$, we must also have $C > B$. Putting all this together we get

$$A > D > C > B.$$

(388) *Problem.* Antonino was rowing his boat upstream. As he passed under a low bridge, his hat was knocked off. It was not, however, until five minutes later that he noticed his hat was missing. At that time he turned around and rowed and rowed with the same effort downstream. He retrieved his hat at a point precisely one kilometre downstream from the offending bridge. How fast was the current flowing?

Solution. There are many ways to approach this problem. The easiest perhaps is to ignore for the moment the movement of the water. Antonino then rows for five minutes after losing his hat, turns around and rows back to where he lost it, which effort should take him again 5 minutes. Now we consider the water as moving and we see that the stream has moved the whole of our consideration downstream for a total elapsed time of 10 minutes. Since one kilometre was covered in this 10 minute interval, we see that the stream was flowing at 6 km per hour.

(389) *Problem.*   Pete the tramp and his dog were five-sixths of the way across a bridge when Pete saw a train coming toward them at 45 mph, Pete turned and ran full speed away from the train. His dog, which was much smarter than Pete, continued to trot toward the oncoming train at a third of Pete's speed and got off the bridge just as the train reached it. Pete got all the way to the other end of the bridge before the train flattened him. How fast did Pete run?

Solution.   Since Pete ran three times as fast as his dog, he had gone 3/6 of the way back across the bridge by the time the dog had trotted the last 1/6 of the way to the closer end of the bridge. When the train got on the bridge, Pete had $5/6 - 3/6 = 2/6$ of the bridge's length left to run. Pete ran the 2/6 of the bridge length in the same time it took the train to cross the entire bridge. Hence the train went three times as fast as Pete ran. Since the train's speed was given as 45 mph, Pete's speed was 15 mph.

(390) *Problem.*   Find all ordered pairs of numbers (x, y) whose sum $x + y$, product $x \cdot y$, and quotient x/y are all equal.

Solution.   Since $xy = x/y$ we see that $xy^2 = x$, which has solutions $x = 0$ (impossible) or $y = \pm 1$. Now set $x + y = xy$. It is clear that $y = 1$ cannot be a solution; but setting $y = -1$ we get $x = \frac{1}{2}$. Thus, the only solution is $(x, y) = \left(\frac{1}{2}, -1\right)$.

(391) *Problem.*   Consider the following "telescoping" series:

$$\sum_{k=1}^{\infty} \frac{1}{(k+1)(k+2)} = \frac{1}{6} + \frac{1}{12} + \frac{1}{20} + \cdots .$$

On the one hand we have

$$\frac{1}{6} + \frac{1}{12} + \frac{1}{20} + \cdots = \left(\frac{1}{2} - \frac{1}{3}\right) + \left(\frac{1}{3} - \frac{1}{4}\right) + \left(\frac{1}{4} - \frac{1}{5}\right) + \cdots = \frac{1}{2};$$

while on the other we get

$$\frac{1}{6} + \frac{1}{12} + \frac{1}{20} + \cdots$$

$$= \left(1 - \frac{5}{6}\right) + \left(\frac{5}{6} - \frac{3}{4}\right) + \left(\frac{3}{4} - \frac{7}{10}\right) + \left(\frac{7}{10} - \frac{2}{3}\right) + \cdots$$

$$= 1.$$

Explain the inconsistency of these two approaches to the problem.

Solution.   The inconsistency becomes apparent when one considers the partial sums:

$$S_n = \sum_{k=1}^{n} \frac{1}{(k+1)(k+2)} = \frac{1}{6} + \frac{1}{12} + \frac{1}{20} + \cdots + \frac{1}{(n+1)(n+2)}.$$

In order for the series to have a sum, the values S_n must have a limiting value, that is,

$$S = \lim_{n \to \infty} S_n = \lim_{n \to \infty} \sum_{k=1}^{n} \frac{1}{(k+1)(k+2)}$$

must exist. In the first approach we see that

$$S = \lim_{n \to \infty} \left[\left(\frac{1}{2} - \frac{1}{3} \right) + \left(\frac{1}{3} - \frac{1}{4} \right) + \cdots + \left(\frac{1}{n+1} - \frac{1}{n+2} \right) \right]$$

$$= \lim_{n \to \infty} \left[\frac{1}{2} - \frac{1}{n+2} \right] = \frac{1}{2} - 0 = \frac{1}{2},$$

while in the second approach we also have

$$S = \lim_{n \to \infty} \left[\left(\frac{4}{4} - \frac{5}{6} \right) + \left(\frac{5}{6} - \frac{6}{8} \right) + \cdots + \left(\frac{k+3}{2k+2} - \frac{k+4}{2k+4} \right) \right]$$

$$= \lim_{n \to \infty} \left[\frac{4}{4} - \frac{k+4}{2k+4} \right] = 1 - \frac{1}{2} = \frac{1}{2}.$$

(392) *Problem.*   A car accelerates uniformly from rest to $8K$ km/h in $K/5$ minutes. It continues at that speed for K minutes, then decelerates uniformly and takes another $K/5$ minutes to come to rest, having traveled exactly $K - 1$ km altogether. The trip took an exact number of minutes. How many?

Solution.   The car can be assumed to travel $4K$ km/h throughout the acceleration and deceleration periods, thus, travelling

$$4K \cdot \frac{K}{5} \cdot \frac{1}{60} = \frac{K^2}{75}$$

kilometres at each end of the trip. In between it goes $8K$ km/h for $K/60$ hours for a total of $2K^2/15$ kilometres in the middle. Altogether it travels

$$\frac{2K^2}{75} + \frac{2K^2}{15} = \frac{12K^2}{75} = \frac{4K^2}{25}$$

kilometres. If this must equal $K - 1$ km, then we have $4K^2 - 25K + 25 = 0$, which factors into $(4K - 5)(K - 5) = 0$. Therefore, the possible solutions for K are 5 and $\frac{5}{4}$. Since the number of minutes for the trip must be an integer, we see that $K = 5$, and the trip took 7 minutes.

(393) *Problem.*   The sequence of positive integers

$$a_1, a_2, \ldots, a_n, \ldots$$

is such that $a_{a_n} + a_n = 2n$ for all $n \geq 1$. Prove that $a_n = n$ for all $n \geq 1$.

Solution. Clearly, $a_i \geq 1$ for $n \geq 1$, since all entries in the sequence are positive integers. The proof will be by induction on n. First, we will examine $n = 1$. We are told that $a_{a_1} + a_1 = 2$. But $a_{a_1} \geq 1$ and $a_1 \geq 1$, which implies that $2 = a_{a_1} + a_1 \geq 2$, forcing $a_{a_1} = a_1 = 1$, and the result is true for $n = 1$.

Next, assume the result is true for all $1 \leq n \leq k$, for some $k \geq 1$. That is, assume that $a_n = n$ for all $1 \leq n \leq k$. Let us now consider $n = k + 1$. If $a_{k+1} = k + 1$, we are done. Therefore, suppose instead that $a_{k+1} \neq k + 1$. Then we have two possibilities, namely $a_{k+1} < k + 1$ or $a_{k+1} > k + 1$. Since a_{k+1} is an integer we have the following cases:

Case (i): $a_{k+1} \leq k$.

Since $a_{k+1} \leq k$, we have by the induction hypothesis that $a_{a_{k+1}} = a_{k+1}$. But then

$$a_{a_{k+1}} + a_{k+1} = 2a_{k+1} \leq 2k < 2(k + 1),$$

which contradicts the given condition that $a_{a_{k+1}} + a_{k+1} = 2(k+1)$.

Case (ii): $a_{k+1} \geq k + 2$.

Let $m = a_{k+1} \geq k+2$. From our given condition we have $a_{a_{k+1}} \leq k$, that is, $a_m \leq k$. By the induction hypothesis we see that $a_{a_m} = a_m$. Since we also have $a_{a_m} + a_m = 2m$, we must have $a_m = a_{a_m} = m$. But this says that

$$k \geq a_{a_{k+1}} = a_m = m = a_{k+1} \geq k + 2,$$

which is again a contradiction.

Therefore, we have $a_{k+1} = k = 1$, and by induction we see that $a_n = n$ for all $n \geq 1$.

(394) *Problem.* Let ABC be a triangle with sides $a = 8$, $b = 7$, and $c = 5$.

(a) Show that the angles A, B, C are three numbers in an arithmetic progression.

(b) Now keeping side a and angle B above fixed, determine all possible integer sided triangles.

Solution. (a) We first observe that three real numbers A, B, C are in arithmetic progression if and only if they are of the form: $x - d$, x, and $x + d$, the sum of which is $3x$. Since $A + B + C = 180°$, we see that the angles are in arithmetic progression if and only if the median angle of the three is $60°$. Thus, we need only show that

$\angle B = 60°$ (it must be the median angle since it is opposite the median side). To do this we invoke the Law of Cosines:

$$b^2 = a^2 + c^2 - 2ac\cos B,$$
$$7^2 = 8^2 + 5^2 - 2 \cdot 8 \cdot 5 \cos B,$$
$$80 \cos B = 64 + 25 - 4940,$$
$$\cos B = \frac{1}{2}.$$

The only positive angle which satisfies this is $\angle B = 60°$, which (by our earlier argument) implies that the angles are in arithmetic progression.

(b) Let us assume that we have an integer sided triangle with $a = 8$ and $\angle B = 60°$. We must find the possible values for a and c. Again we will use the Law of Cosines:

$$b^2 = 8^2 + c^2 - 2 \cdot 8c \cos 60° = 64 + c^2 - 8c$$
$$b^2 - 48 = c^2 - 8c + 16 = (c - 4)^2.$$

Thus, $b^2 - 48 = k^2$, a perfect square, where $k = |c - 4|$. Then

$$b^2 - k^2 = 48,$$
$$(b - k)(b + k) = 48$$
$$= 2^4 \cdot 3.$$

Clearly $b - k < b + k$. Since their sum is even we note that $b - k$ and $b + k$ are both even or both odd; since their product is 48, an even number, we conclude that both $b - k$ and $b + k$ are even. Thus, we need to factor 48 into two even factors: $b - k$ and $b + k$, and then solve for b and k:

$b - k$	$b + k$	b	k	c
2	24	13	11	15
4	12	8	4	8
6	8	7	1	5 or 3

Thus, there four such integer sided triangles: $(8, 13, 15)$, $(8, 8, 8)$, $(8, 7, 5)$, and $(8, 7, 5)$.

(395) *Problem.*   Seventy-five cows have grazed all the grass in a 60-acre pasture in 12 days, and eighty-one cows have grazed all the grass in a 72-acre pasture in 15 days.

How many cows can graze all the grass in a 96-acre pasture in 18 days?

Solution. The above problem is a special case of the following problem, which appeared in Newton's *Arithmaetica Universalis* (1707):

a cows graze *b* acres bare in *c* days,
a' cows graze *b'* acres bare in *c'* days,
a'' cows graze *b''* acres bare in *c''* days;
what relation exists between the nine magnitudes *a* to *c''*?

We will solve this general problem first.

The problem can be solved only if certain reasonable assumptions are made. We assume that each cow consumes a constant amount of grass per day, that the initial amount of grass in each acre is a constant r, and that the daily growth rate of the grass in each acre is a constant s.

In the first case, the total amount of grass initially present in the pasture is br, that the amount of grass which grows during the period of grazing is bcs, and the cows consume an amount of grass equal to ca. Thus,

$$br + bcs - ca = 0.$$

Similarly,

$$b'r + b'c's - c'a' = 0,$$
$$b''r + b''c''s - c''a'' = 0.$$

Now consider the homogeneous linear system of equations:

$$bx + bcy - caz = 0,$$
$$b'x + b'c'y - c'a'z = 0,$$
$$b''x + b''c''y - c''a''z = 0. \tag{1}$$

Clearly, this system has the non-zero solution $(x, y, z) = (r, s, -1)$. Thus, the determinant of the system must be 0. That is,

$$\begin{vmatrix} a & bc & ca \\ b' & b'c' & c'a' \\ b'' & b''c'' & c''a'' \end{vmatrix} = 0,$$

and this is the required relation. If any eight of the nine quantities *a* to *c''* are known, the ninth can be determined.

In the current problem, we have

$$a = 75,\ a' = 81,\ b = 60,\ b' = 72,\ b'' = 96,\ c = 12,\ c' = 15,\ c'' = 18.$$

Relation (1) then yields $a'' = 100$.

(396) *Problem.*   Start with four containers that contain known quantities of golf balls. A *legal move* is defined as removing one ball from each of three of the containers and placing these three balls in the fourth container. For example, if the four containers contained 10, 12, 14, and 16 balls, then the result of one of the four legal moves available to us would be 13, 11, 13, and 15 balls, respectively.

Determine a condition or a set of conditions on the initial distribution of golf balls such that there is a sequence of legal moves which will result in all the containers having the same number of balls.

Solution.   Let a, b, c, d be the initial number of golf balls in containers 1, 2, 3, 4, respectively. Without loss of generality, we assume that $a \geq b \geq c \geq d$. Let x be the number of legal moves which increase the number of golf balls in the first container, let y, z, and w be similarly defined for the second, third, and fourth containers, respectively. Let T be the total number of golf balls in use. Then $T = a + b + c + d$.

Clearly, if there are two empty containers, then there are no legal moves available, so we will insist that there is at most one container having no golf balls.

We are seeking values of x, y, z, and w such that after applying this number of legal moves of each of the four allowable types, we end up with $T/4$ golf balls in each container. Thus, as a first observation, we see that T must be a multiple of 4 in order that we have any chance of finding a solution. We need to find a solution to the following system of linear equations:

$$a + 3x - y - z - w = \frac{1}{4}T,$$

$$b - x + 3y - z - w = \frac{1}{4}T,$$

$$c - x - y + 3z - w = \frac{1}{4}T,$$

$$d - x - y - z + 3w = \frac{1}{4}T,$$

which, since $T = a + b + c + d$, can be rewritten as

$$3x - y - z - w = -\frac{3}{4}a + \frac{1}{4}b + \frac{1}{4}c + \frac{1}{4}d,$$

$$-x + 3y - z - w = \frac{1}{4}a - \frac{3}{4}b + \frac{1}{4}c + \frac{1}{4}d,$$

$$-x - y + 3z - w = \frac{1}{4}a + \frac{1}{4}b - \frac{3}{4}c + \frac{1}{4}d,$$

$$-x - y - z + 3w = \frac{1}{4}a + \frac{1}{4}b + \frac{1}{4}c - \frac{3}{4}d.$$

Using Gaussian Elimination, we see that this system is equivalent to

$$x - w = \frac{1}{4}(d - a),$$

$$y - w = \frac{1}{4}(d - b),$$

$$z - w = \frac{1}{4}(d - c).$$

It is clear that this system has a solution in integers if and only if $d - a$, $d - b$, and $d - c$ are all multiples of 4. It is easy to see that this is equivalent to stating that a, b, c, d must all have the same remainder on division by 4. Since d was a maximal element from the set $\{a, b, c, d\}$, we see that the right hand sides of the above system are all non-negative. Thus, we may choose $w = 0$, which yields values for x, y, z which are all non-negative, and this will produce a solution.

Therefore, the initial distribution of values requires that the values of a, b, c, and d all have the same remainder on division by 4 and that at most one container is empty, and this necessary condition is also sufficient since we can then set $w = 0$ and obtain non-negative integer solutions for x, y, and z. In particular, the example in the problem cannot be solved since the initial distribution has some values which leave a remainder of 0 and some which leave a remainder of 2 on division by 4. However, for example, the initial distribution 9, 9, 13, 17 can be solved: $x = y = 2$, $z = 1$, and $w = 0$.

Alternate Approach (by Peter Smoczynski): First observe that after one legal move the contents of one container has increased by 3 balls while the contents of each of the others has decreased by 1. This means that the differences between pairs of containers either stays the same or changes by 4. Thus, after any number of legal moves the set of initial differences can only be altered by multiples of 4. Thus, in order to get the set of differences to be all zeroes, we must start with the set of differences as multiples of 4. If we always apply the rule that we will increase the contents of one of the containers with the least amount in each round, then the greatest difference will never increase, and indeed will decrease by 4 unless there were two or more containers with the same least amount. Since such an operation will reduce the number of balls with least amount for the next round, a sequence of these operations must ultimately reduce the maximum difference until we reach 0. We again must exercise caution that we do not start with more than one empty container.

(397) *Problem.* At a party there are 201 people of five different nationalities. In each group of six at least two people have the same age. Show that there are at least five people of the same nationality, of the same age, and of the same sex at the party.

Solution. Clearly there is at least one nationality to which at least 41 people belong. Among these 41 there are at least 21 who also are the same sex. Clearly there can be at most five different ages at the party or we could select a group of six, each from a different age, which contradicts the problem statement. Thus, among the 21 people who have the same nationality and the same sex, there must be at least five who also have the same age.

(398) *Problem.* A difficult mathematical competition consisted of a Part I and a Part II with a combined total of 28 problems. Each contestant solved exactly seven problems altogether. For each pair of problems, there were exactly two contestants who solved both of them. Prove that, if every contestant solved at least one problem in Part I, then at least one contestant solved at most three problems in Part II.

Solution. Consider an arbitrary problem and let it be solved by exactly r contestants. These contestants together solve $6r$ other problems (counting multiplicities). Each of the remaining 27 problems is counted twice in $6r$ (since each pair of problems is solved by exactly 2 contestants), so that $6r = 2 \times 27$ or $r = 9$. It follows that each problem is solved by exactly 9 contestants, and the number of contestants is $9 \times 28/7 = 36$.

Suppose that every contestant solves one, two, or three problems in Part I. Let n be the number of problems in Part I, and x, y and z be the respective numbers of contestants who solve one, two, and three of these problems. Then

$$x + y + z = 36, \tag{1}$$
$$x + 2y + 3z = 9n, \tag{2}$$
$$y + 3z = 2\binom{n}{2}. \tag{3}$$

The last equation arises from the observation that, for every pair of problems in Part I, exactly 2 contestants solve both of them. Multiplying (1), (2), and (3) by -3, 3, and -2, respectively, and adding the resulting equations, we have

$$y = -2n^2 + 29n - 108 = -2\left(n - \frac{29}{4}\right)^2 - \frac{23}{8} < 0,$$

an impossible situation. Hence, at least one contestant solves 4 or more problems in Part I, and consequently 3 or fewer problems in Part II.

(399) *Problem.*   Suppose that r is a non-negative rational taken as an approximation to $\sqrt{2}$. Show that $\frac{r+2}{r+1}$ is always a better rational approximation.

Solution.   First, we observe that if r is rational then so also is $\frac{r+2}{r+1}$. We must simply show that

$$\left| \frac{r+2}{r+1} - \sqrt{2} \right| < \left| r - \sqrt{2} \right|.$$

The following argument proves this inequality:

$$\left| \frac{r+2}{r+1} - \sqrt{2} \right| = \left| \frac{(r+2) - \sqrt{2}(r+1)}{r+1} \right|$$
$$= \left| \frac{r(1 - \sqrt{2}) + 2 - \sqrt{2}}{r+1} \right|$$
$$= \left| \frac{r(1 - \sqrt{2}) + \sqrt{2}(\sqrt{2} - 1)}{r+1} \right|$$
$$= \frac{(\sqrt{2} - 1)| r - \sqrt{2} |}{r+1}$$
$$\leq (\sqrt{2} - 1)| r - \sqrt{2} |$$
$$< | r - \sqrt{2} |.$$

(400) *Problem.*   Let a_1, a_2, a_3, a_4, a_5 be a five-term geometric sequence satisfying the inequality $0 < a_1 < a_2 < a_3 < a_4 < a_5 < 100$, where each term is an integer. How many of these five-term geometric sequences are there? (For example, the sequence $(3, 6, 12, 24, 48)$ is a sequence of this type.)

Solution.   Let $\frac{n}{m}$ be the common ratio of the geometric sequence, where n and m are relatively pnme integers (that is, they have no common factor larger than 1), with $n > m$. Now $a_5 = a_1 \times \frac{n^4}{m^4}$, so that we let $a_1 = km^4$ where k is a positive integer. Thus, our geometric sequence becomes

$$km^4, km^3 n, km^2 n^2, kmn^3, kn^4,$$

where $a_5 = kn^4 < 100$. If $n \geq 4$, then $kn^4 \geq n^4 \geq 256 > 100$. Thus, $n \leq 3$. Hence, there are three cases to consider:

   (i) $n = 3$, $m = 2$. Then $a_5 = 81k < 100$, so that $k = 1$. The only solution is $(16, 24, 36, 54, 81)$.

   (ii) $n = 3$, $m = 1$. Again $a_5 = 81k < 100$, so that $k = 1$. The only solution is $(1, 3, 9, 27, 81)$.

   (iii) $n = 2$, $m = 1$.   Then $a_5 = 16k < 100$, so that $1 \leq k \leq 6$. This gives us six solutions, one for each value of k: $(1, 2, 4, 8, 16)$, $(2, 4, 8, 16, 32)$, $(3, 6, 12, 24, 48)$, $(4, 8, 16, 32, 64)$, $(5, 10, 20, 40, 80)$, and $(6, 12, 24, 48, 96)$.

Thus, there are a total of eight such five-term sequences within the specified range.

(401) *Problem.*   Several people go to a pizza restaurant. Each person who is "hungry" wants to eat either 6 or 7 slices of pizza. Everyone else wants to eat only 2 or 3 slices of pizza. Each pizza at the restaurant has 12 slices. It turns out that four pizzas are not sufficient to satisfy everyone, but that with 5 pizzas there would be some pizza left over. How many people went to the restaurant, and how many of these were "hungry"?

Solution.   Let x be the number of "hungry" people, and let y be the number of non-"hungry" people.

We know that $6x + 2y > 4 \times 12 = 48$, and $7x + 3y < 5 \times 12 = 60$. Because x and y are integers, the quantities $6x + 2y$ and $7x + 3y$ are also integers and the first of these must be even. Therefore, we can rewrite the inequalities as:

$$6x + 2y \geq 50 \qquad \text{and} \qquad 7x + 3y \leq 59.$$

Taking the difference we see that $x + y$, the total number of people, is at most 9. If x is seven or less, then

$$6x + 2y \leq 6x + 2(9 - x) = 4x + 18 \leq 4 \times 7 + 18 = 46 < 50,$$

which is impossible, so that x is either 8 or 9. On the other hand, if $x = 9$, we have $7x = 63 > 60$, also impossible. Thus, $x = 8$. If $y = 0$, the four pizzas would suffice ($8 \times 6 = 48 = 4 \times 12$). Thus, $y \geq 1$. Since we know there at most 9 people altogether, we see that $y = 1$. Therefore, there are 8 "hungry" people and 1 non-"hungry" person.

(402) *Problem.* $P(x)$ is a polynomial of degree greater than 2 with integer coefficients such that $P(2) = 13$ and $P(10) = 5$. It is known that P has a root which is an integer. Find this root.

Solution.   Let r be an integer root of the polynomial P. Then $P(x) = (x - r)Q(x)$, where $Q(x)$ is a polynomial of degree 1 less than that of $P(x)$, also with integer coefficients. The two given conditions then become:

$$13 = P(2) = (2 - r)Q(2),$$
$$5 = P(10) = (10 - r)Q(10).$$

Since $Q(x)$ has integer coefficients, both $Q(2)$ and $Q(10)$ are integers, as are $2 - r$ and $10 - r$ (since r is an integer). Thus, the two equations above are integer equations, and we conclude that $2 - r$ evenly divides 13, and $10 - r$ evenly divides 5.

This implies that $2 - r \in \{\pm 1, \pm 13\}$ (or $r \in \{1, 3, -11, 15\}$) and that $10 - r \in \{\pm 1, \pm 5\}$ (or $r \in \{9, 11, 5, 15\}$). The only value in common to both is $r = 15$, and that is the solution.

Note: the fact that the degree was greater than 2 was never used. Indeed, the solution $P(x) = 15 - x$; for all x satisfies the problem statement with P a linear polynomial. However, we can deduce from the solution that the degree of P must not be 2, for if it were 2, then $Q(x)$ would be a linear polynomial satisfying $Q(2) = Q(10) = -1$ and thus, Q would be the constant polynomial and P would be linear.

(403) *Problem.*   A line ℓ with slope $m = 2$ cuts the parabola $y^2 = 8x$ to form a chord. Find the equation of ℓ if the mid-point of the chord lies on $x = 4$.

Solution.   Since the slope is 2, the line l has equation $y = 2x + b$. We must only find the value of b. Inserting this value of y into the equation of the parabola yields:

$$(2x + b)^2 = 8x,$$
$$4x^2 + 4bx + b^2 = 8x,$$
$$4x^2 + 4(b - 2)x + b^2 = 0,$$
$$x = \frac{4(2 - b) \pm \sqrt{16(b^2 - 4b + 4) - 16b^2}}{8},$$
$$= 1 - \frac{b}{2} \pm \sqrt{1 - b}.$$

The two roots are $x_1 = 1 - \frac{1}{2}b - \sqrt{1 - b}$ and $x_2 = 1 - \frac{1}{2}b + \sqrt{1 - b}$. These are the coordinates of the end-points of the chord. Since we are looking for the mid-point of the chord to have x-coordinate 4, we must have $(x_1 + x_2)/2 = 4$. Thus, $(2 - b)/2 = 4$, which yields $b = -6$. Therefore, the line ℓ has equation $y = 2x - 6$.

(404) *Problem.*   Show that three tangents can be drawn from the origin to the curve given by

$$y = x^3 - 13x^2 + 10x - 36.$$

Solution.   Let $P(a, a^3 - 13a^2 + 10a - 36)$ be any point on the given curve. The tangent to the curve at this point has slope $m = 3a^2 - 26a + 10$, which means that the tangent line to the curve at the point P has equation:

$$y - (a^3 - 13a^2 + 10a - 36) = (3a^2 - 26a + 10)(x - a)$$
$$\text{or} \quad y = (3a^2 - 26a + 10)x - 2a^3 + 13a^2 - 36.$$

If the tangent line is to pass through the origin, we must then have $2a^3 - 13a^2 + 36 = 0$. This factors thus:

$$2a^3 - 13a^2 + 36 = (a - 2)(a - 6)(2a + 3).$$

Thus, the only values of a which work are 2, 6, and $-\frac{2}{3}$. It can be checked that the lines $y = -30x$, $y = -38x$, and $y = \frac{223}{4}x$ are tangents to the curve at the respective points $(2, -60)$, $(6, -228)$, and $(-\frac{2}{3}, -\frac{669}{8})$.

(405) *Problem.*   A man sold a house for \$75000 and gained a certain per cent on the original cost. If the cost had been $16\frac{2}{3}\%$ less, his gain would have been 25% greater. Find the original cost of the house.

Solution. Let C be the original cost of the house. The profit on the house is thus, $\$75000 - C$. If the cost was $16\frac{2}{3}\%$ less, then the cost would be $\frac{5}{6}C$, and the profit would be $\$75000 - \frac{5}{6}C$. This represents an increase in profit of $\frac{1}{6}C$. As a percentage we then have

$$\frac{\frac{1}{6}C}{75000 - c} = 0.25,$$

$$\frac{1}{6}C = 18750 - \frac{1}{4}C,$$

$$\frac{5}{12}C = 18750,$$

$$C = 45000.$$

Thus, the original cost of the house was $\$45000$. Let us check. The gain made on the sale was $\$30000$. If the cost had been $16\frac{2}{3}\%$ less (that is, if the cost had been $\$37500$), his gain would also have been $\$37500$, representing a 25% increase in his gain (up $\$7500$ from $\$30000$).

(406) *Problem.* Jack and Jill run 10 kilometres. They start at the same point, run 5 kilometres up a hill, and return to the starting point by the same route. Jack has a 10-minute head start and runs at the rate of 15 km/hr uphill and 20 km/hr downhill. Jill runs at the rate of 16 km/hr uphill and 22 km/hr downhill. How far from the top of the hill are they when they pass going in opposite directions?

Solution. When Jill starts to run, Jack has already covered 2.5 km uphill. Thus after 20 minutes of running Jack will have reached the top of the hill. At this time Jill will have only run 10 minutes and will have covered $\frac{8}{3}$ km. This leaves $\frac{7}{3}$ km between the two runners, when Jack turns around to run downhill. When the two runners are running toward each other, the distance between them is closing at the (combined) rate of 36 km/hr. At that rate it takes $\frac{35}{9}$ minutes to cover the remaining $\frac{7}{3}$ kilometres. During this time Jack will have run $\frac{35}{27}$ km back down the hill. Thus, they are $\frac{35}{27}$ km from the top of the hill when they meet going in opposite directions.